SpringerBriefs in Climate Studies

SpringerBriefs in Climate Studies present concise summaries of cutting-edge research and practical applications. The series focuses on interdisciplinary aspects of Climate Science, including regional climate, climate monitoring and modeling, palaeoclimatology, as well as vulnerability, mitigation and adaptation to climate change. Featuring compact volumes of 50 to 125 pages (approx. 20,000–70,000 words), the series covers a range of content from professional to academic such as: a timely reports of state-of-the art analytical techniques, literature reviews, in-depth case studies, bridges between new research results, snapshots of hot and/or emerging topics.

Author Benefits: SpringerBriefs in Climate Studies allow authors to present their ideas and readers to absorb them with minimal time investment. Books in this series will be published as part of Springer's eBook collection, with millions of users worldwide. In addition, Briefs will be available for individual print and electronic purchase. SpringerBriefs books are characterized by fast, global electronic dissemination and standard publishing contracts. Books in the program will benefit from easy-to-use manuscript preparation and formatting guidelines, and expedited production schedules. Both solicited and unsolicited manuscripts are considered for publication in this series. Projects will be submitted to editorial review by editorial advisory boards and/or publishing editors. For a proposal document, please contact the Publisher.

Gerrit Lohmann

Conversations on Climate: The People Behind the Science

Springer

Gerrit Lohmann
Department of Climate Science
Alfred Wegener Institute
Bremerhaven, Bremen, Germany

ISSN 2213-784X ISSN 2213-7858 (electronic)
SpringerBriefs in Climate Studies
ISBN 978-3-032-19146-5 ISBN 978-3-032-19147-2 (eBook)
https://doi.org/10.1007/978-3-032-19147-2

This work was supported by Alfred Wegener Institute Helmholtz Centre for Polar and Marine Research.

This Springer imprint is published by the registered company Springer Nature Switzerland AG
The registered company address is: Gewerbestrasse 11, 6330 Cham, Switzerland

Preface I

Anyone who studies the climate quickly realizes what a miracle our planet is. This tiny little spaceship manages to house billions of species. In the middle of cold and empty space, the Earth is a place full of warmth, light, water, and nutrients. But all these things cannot be taken for granted and are in a unique balance with each other for life to thrive. It is all the more dangerous when this balance, which has evolved over millions of years, is disturbed by our way of life.

Climate has always changed, but mostly without human influence and on much longer time scales. In the past, there were times when our planet was almost completely ice-free, but also times when kilometer-thick ice made part of the Earth a hostile place to live. Also, concentrations of greenhouse gases like CO_2 were not always stable in the atmosphere: in the deep past, much of it was temporarily stored in plants. Over time, parts of it became petroleum or hard coal, safely locked away under the surface. Mankind now reintroduces it into the atmosphere, burning it to generate energy. The greenhouse gases emitted this way imbalance the Earth's heat, less heat escapes from our planet, while the downward shortwave radiation from the Sun is nearly unhindered. The result: global warming.

Now, in just a few decades, human emissions have considerably affected the Earth's climate, with ecological, economic, and social consequences already in place. Some of these changes may be irreversible. Weather extremes are becoming more frequent and intense, flora and fauna are exposed to droughts and floods. The oceans, which serve as one of our most valuable natural carbon stores, experience severe acidification. Crop failures are expected to significantly affect the agricultural sector and may increase famines. Sea levels are rising due to higher temperatures and in the future due to the expected thawing of ice sheets. The consequences of climate change have visible negative impacts on the functioning of the natural system with the potential to destroy cities, cultures, and communities around the world. People can lose their homes, and some even their lives.

This shows that climate change is one of the greatest challenges of our time and has become more urgent than ever. In response to the many challenges, the global community has formulated Sustainable Development Goals and an action plan for the future that includes economic, ecological as well as social aspects. We must find sustainable solutions to protect our planet—our future. Scientists and civil society want to contribute to these challenges. This book gives an insight into the research and motivation of scientists, the key questions they want to solve, and their ideas on how to do so. We hope it inspires you and hope you enjoy reading it.

Bremerhaven, Germany
July 2023

Gerrit Lohmann

Preface II

The consequences of climate change have become increasingly visible and are now regularly discussed in political and societal debates. While climate scientists have been warning about the potential consequences of anthropogenic greenhouse gas emissions for decades, their call for action has only recently been given the attention and urgency it deserves. At the same time, the field of climate science has grown significantly and has become more interdisciplinary. As a result of more sophisticated technologies, higher data availability, and more advanced climate models, the fundamental message is stronger than ever: the impact of human activity on the Earth's climate is clear, and severe and irreversible consequences are already occurring all over the world. Although it will take time for the climate to stabilize again, strong reductions in greenhouse gas emissions can still limit the level of warming and its detrimental consequences.

We, the authors of this book, are a group of students from different disciplines within the natural sciences, connected through a common interest in climate research. Through a biannual scientific seminar organized by the German Academic Scholarship Foundation, we were given the opportunity to interview leading climate scientists from different disciplines including theoretical physics, paleo-climatology, and oceanography. With a main focus on the insights provided by climate models, the interviews cover topics ranging from causes and consequences of climate change, the role of the biosphere, tipping elements, and attribution science. Additionally, the interviews address the political dimension of climate science, including reflections on post-colonialism and feminist perspectives on climate change. As students who aspire to make a meaningful contribution to the necessary societal transformation towards sustainability, we were eager to learn what it means to work as a climate scientist and to get in touch with the people who shape this field. Under the direction of climate physicist Gerrit Lohmann of the Alfred-Wegener-Institute in Bremerhaven, Germany, we envisioned a book that reflected the diversity of climate science. We wanted to not only present research findings but also to give an impression of the people behind the science. With the selection of interviews, we try to cover different aspects of climate research and represent people with different backgrounds.

Reading our book will inform you of the major findings of renowned climate scientists, their opinions about current challenges, and what they expect from their field in the future. In addition, you can look forward to personal reflections about the daily lives, motivations, and future aspirations of different researchers and learn how they have experienced the development of climate science over the past decades. In this way, these conversations can also serve as inspiration for students who consider dedicating their professional lives to climate science. The interviews showcase how many diverse pathways can lead to climate research and which questions remain to be answered.

More generally, this book is meant for anyone who feels concerned with the challenges associated with climate change and strives to contribute to the solutions we need. It provides a look behind the scenes and shares personal insights on how complex and also frustrating climate science can be, but, most importantly, it demonstrates what keeps climate scientists motivated and hopeful and how exciting and inspiring working in this field can be.

Bremerhaven, Germany The Authors
July 2023

Acknowledgements

A heartfelt thank you to Anne Wildfeuer, Lukas von Maltitz, Alexandra Haseloff, and Stephanie Richters for organizing the scientific college of the German Academic Scholarship Foundation, which allowed us to come and work together as a group and therefore served as the basis for this project in the first place. We sincerely acknowledge their efforts to support our learning in the best way possible at all times, particularly when we were forced to amend the plans for our first working phase due to the COVID-19 pandemic. We highly appreciated the opportunity to have an additional meeting in Bremerhaven, which included a visit to the laboratories of ice and sediment cores in the Alfred-Wegener-Institute as well as the Klimahaus.

We wish to express our gratitude to all the climate scientists who agreed to be interviewed for this book. Thank you for sharing your insights and honest reflections with us.

Our deepest gratitude goes to our supervisor, Gerrit Lohmann, for guiding us through this big project, for providing inspiration and assistance at all stages of our work, and for his openness to our ideas. We also want to extend our thanks to his colleague Martin Werner for supporting us during the initial phase of the project and to Christian Stepanek for his input and regular exchange. Finally, without the Alfred-Wegener-Institute's generous financial support, this book could not have been published under an open access license.

Contents

Chapter 1
Reto Knutti: "The Facts Don't Speak for Themselves"

Lina Bernert, Lukas Schmitt, Marius Schulz, and Moritz Thies

©Manuel Rickenbacher

Reto Knutti (*1973) is a professor at the Department of Environmental Systems Science at the Swiss Federal Institute of Technology in Zurich (ETH Zurich). He studied physics at the University of Bern and did his Ph.D. on the probability and predictability of future climates. As a postdoc, he was already a lead author for the reports of the Intergovernmental Panel on Climate Change (IPCC). He is known for his scientific contributions to climate modeling, while his other research focuses include the attribution of observed changes to human activities, and extreme weather

Date of the Interview: February 13, 2023

G. Lohmann, *Conversations on Climate: The People Behind the Science*,
SpringerBriefs in Climate Studies, https://doi.org/10.1007/978-3-032-19147-2_1

events. In addition, he is strongly involved in climate communication and policy advice.

What do you enjoy most about your work?
Many things. We have several missions: we educate people, we do research, and I do a lot of outreach, including policy dialogue. In all of these areas, there are things that I enjoy. For example, I enjoy seeing young people who are fascinated by science. They develop from a kind of please-feed-me mentality into independent researchers and leaders and are ready to take on responsibility in our society. In research, I enjoy solving puzzling problems, even though you do it less and less as you get older. Another part I enjoy is trying to put the facts and figures into a form where people understand them, find them interesting, and that then ideally can trigger something inside them.

How did you imagine your professional future when you were a student?
I didn't imagine my professional future. Unlike others who knew from the beginning that they ultimately wanted to be scientists, I never saw that as the only goal. I studied physics because I was interested in it, but I didn't know where it would lead me. And then, during my thesis, I thought I wanted to do something relevant to society.

The first UN climate reports in the 1990s were not necessarily on my radar. Back then, you also couldn't specialize as much in your studies as you do today. That only happened in my diploma thesis. By coincidence, there was a thesis on climate modeling advertised, which caught my interest because I grew up in the mountains and can relate to nature, and it was something societally relevant.

In my doctoral thesis, I delved further into the climate direction and later went abroad also to see another part of the world. During that time, I was at the National Center for Atmospheric Research in Boulder, Colorado, where many hundred people are working on atmospheric and climate research. From chemistry to weather to measurements to modeling, it is almost impossible to find a better place. At that time, I never anticipated that my application to ETH two and a half years later would be successful and that I would stay at a university.

Were there any important moments along the way?
Yes, I was lucky to work for an outstanding group in Bern. My thesis advisor Thomas Stocker (Professor of Climate and Environmental Physics, Physics Institute, University of Bern, Switzerland) was involved as a lead author for the IPCC. That's why my work also quickly got a lot of visibility, and I became a lead author of the IPCC reports when I was still a postdoc, so I was much younger than you normally are. All of that leads to a growing network.

What did you learn in the US? Is there a different mentality in the research there?
In research, the US is not much different from Europe. Both are very international. They are very similar if you compare them, for example, to Asia, where research is a little different because of the hierarchical system. There, people in lower positions are simply told what to work on, and they work incredibly diligently and efficiently

on it. In Asia, you can't question your boss. In the US, on the other hand, it's perfectly fine as a master's student to tell your professor that something doesn't make sense.

Is international experience important in terms of a career in science?
I think it is. It's not that the ways of working are radically different; it's more about the way you approach a scientific question. In Bern, I was in a group that mainly worked with paleoclimatology, reconstructions, and ice cores. Then, over time, you think like your supervisor, and the world becomes an energy balance question. Where I worked in the US, everyone was thinking about high-resolution and spatial patterns like El Niño, so the focus was completely different. In dynamics, you look at everything from a potential vorticity perspective; at the Potsdam Institute for Climate Impact Research, tipping points and planetary boundaries are important; oceanographers only see underwater; and then there are the chemists. The world is complicated, and you have to leave out a lot when describing it. Depending on what you leave out, you come up with different perspectives.

What is your research approach?
Over time, you take a broader view, especially when you go abroad. Personally, I mainly think about climate processes, forcings, feedbacks, and climate sensitivity. Another question I am looking into is how to deal with uncertainties and risks. Together with Erich Fischer, we do a lot of research on extreme weather and climate events like heat waves or heavy precipitation. Another thing we do is to develop climate scenarios for Switzerland on a local level.

You mentioned earlier that you grew up very close to nature in the mountains. How did that influence you?
I originally grew up in Gstaad, skiing from a very young age. At the age of 15, I started to climb, and do mountain and glacier tours. These sports made me not only being close to the elements, they also fascinated me scientifically, and I developed a kind of basic connection to nature. In addition, I spent several summers in the Alps with my uncle's cows, which all had a strong influence on me.

Are you still spending a lot of time in nature?
Yes, as much as I can with two children. I'm not climbing the Alps as high as I used to, but we often stay in huts without electricity or running water.

How does it affect you when you see the glaciers melting and the dwindling amount of snow in winter?
It's hard to say. You see it happening, and you tell yourself, "if I can contribute here to understand this better, then maybe we have a chance to deal with it." That's the rational aspect. But then there is also the emotional discussion. Look at this picture, for example [takes out his smartphone]. I learned to ski on this slope. Today, there is just rubble where you used to walk through snow. That triggers something in me, even if it isn't immediately related to my daily work. I try not to let it influence my work or communication of the severity of climate change, but sometimes I feel like the world is ending. Even though I became a climate scientist not as a Greenpeace activist but as a physicist, it does bother me.

Do you sometimes have doubts about your work?
It was the right decision for me to do this work. After 25 years of climate research, I still enjoy it, and there is still a lot to do. However, when I look at how society is developing, I sometimes wonder why I'm doing this at all. We produce facts and figures; the simulations get better and better, but little change follows. And then I think: either people don't understand, or they don't want to hear it, or they don't care, or they don't know how to respond. But then we could also stop the research. My claim is not that research can decide what needs to be done—that is a societal and political decision—but the research must at least be a starting point for the discussions. If the facts don't matter and everyone can construct alternative facts and their own reality, you must ask yourself whether science is still needed. That's where I sometimes get a bit frustrated.

Is there a research result or outcome that you are particularly proud of?
Proud is maybe the wrong expression; I am not a person who is proud. But some things have been particularly fun for me. For example, in 2009, with Susan Solomon, we showed that most of the climate change is irreversible. Even if we go to net-zero emissions today, we're essentially stuck with the climate change we have now. That has fundamental implications for how we manage these risks. On many other issues, like air pollution and water quality, you can fix it when it goes wrong. Here, we're not going to fix it anymore.

With Sebastian Sippel and Erich Fischer, we also used machine learning methods to explore the human impact of extreme weather events and climate change. We were able to detect climate change in every single day of global weather—not in a single place, but if you look at the world as a whole. Since 2012, we can detect climate change every single day based on spatial patterns. Those are the kinds of things that I enjoyed. I think it's in the minds of most scientists not to be proud of the past. You're always thinking about the next project and future questions.

You said that climate change is seen based on spatial patterns. So it is not just based on the energy that the Earth system contains?
No. You can subtract the mean value from the world map and still see climate change. That's because the land is warming more than the ocean, the high latitudes are warming more than the tropics, and many other smaller patterns also exist. The question is simply where to look. Then, in hindsight, the results always seem trivial.

What has happened in climate models since you did your Ph.D. in 2002?
A revolution. We have a much greater understanding of how the Earth system works, partly because we now have 20 years of targeted observations. You always have to think about the fact that many observations were completely unsystematic until the 1980s. Nobody thought about climate change. With the ice sheets, they didn't start taking data until the 2000s. Now that we have all the satellite data, we have a better understanding of the processes and can also better construct the models.

Even more important is the revolution in computing. When I started, the whole climate and environmental physics group, about 40 people, had about 28 GB of storage. Today, a PowerPoint presentation already needs more than half a Gigabyte.

Then there are all the programming tools. In the past, there was only Fortran. If you wanted to do a figure, there was the NCAR Graphics Library in Fortran. For a simple line plot, you needed a Fortran routine with 500 lines. Today, that's only one line. Many ideas already existed back then, but they weren't technically possible. Perhaps it's hard to grasp all the things that have changed with digitalization, including access to data and literature. I don't feel old, but we used to go to the library every Friday, browse through the *Journal of Physical Oceanography*, then go to the copier and copy interesting articles.

Were climate models easier to understand in the past because they were less complex, or are they easier to understand today because the results can be presented better?
There is still a spectrum of different models. It isn't the case that only complex, high-resolution models are used today because of the greater computing capacity. For specific questions, people still look at simple energy balance models. They're not much different today than what Gerald North and Tom Wigley did 30 years ago. The simple models are not replaced, but more has become possible with the complex ones, and they match reality better and better in many areas. The downside to complex models is that they are very difficult to understand because they have an incredible number of degrees of freedom. There is really no one who understands every part of a climate model anymore.

Do you think that's a problem?
Not necessarily, but you have to create structures that catch that. There used to be one person in an institution who kept all the code together. The mastermind behind the code doesn't exist anymore. This requires much more coordination and agreement, which limits creative play. Many software components are no longer programmed by domain scientists but by computer scientists. The whole numerical part is isolated from the physics.

Are the lack of observations the main limitation of climate models?
A lack of observations was a problem until 1990 or 2000. Today, we have extremely good observations in many areas. However, long-term observations like precipitation data before 1950 are missing, and we can't go back in time and measure again.

In terms of high-resolution models, we also need to distinguish between process complexity and high resolution. Process complexity means understanding the individual factors, such as permafrost and ice sheets. The other is simply a higher spatial resolution to represent the dynamics better. For high-resolution models, the limitation is mainly a technical issue. There is still a factor of 100 or so missing in computing for a global climate simulation at the kilometer resolution. Although it certainly helps, I don't think that having more computing power alone will solve all problems.

What are you working on right now to improve modeling?
A lot of things. We are trying to simulate extreme scenarios such as heat waves, heavy precipitation, or droughts in the context of national and European climate scenarios as well as their impacts. In addition, we are working on model evaluation, which means that we are investigating which model is suitable for which problem.

We are also working on the so-called pattern effect, which describes how the Pacific Ocean is warming differently than the models predicted. The question is whether this deviation is random or systematic.

In 2009, you wrote an editorial comment called "The End of Model Democracy," where you state that while the accuracy of models is improving, more and more aspects are also being taken into account, and thus the uncertainties remain the same. Is that still true, or has it changed since then?
For the range of models, it is indeed the case that the uncertainties have barely decreased over a long time. For example, we used to model the whole land surface as a parking lot. Today, we have complex land models with grass, different types of trees, and land use. This significantly increases the range of model behavior.

However, for specific quantities, we also reach the point where we reduce uncertainties. For instance, the uncertainty of predicted temperature scenarios by the end of the century is smaller today than it was ten years ago because we can exclude certain extreme cases. However, for other variables such as drought or precipitation, the uncertainties are still relatively large.

In the IPCC report, uncertainties are often calculated using energy balance models. What are the strengths of these simple models?
On the one hand, they are conceptually very simple to understand: the difference between incoming and outgoing energy results in a temperature increase, which is the basic conservation of energy. In addition, the number of uncertain quantities, such as radiative forcing, feedback effects, or albedo, is limited, so you can systematically determine the uncertainties. This is impossible for complex models due to their limited computational capacity and the amount of data.

Moreover, you can calibrate the simple model to the complex model and then calculate an emission scenario in a fraction of a second. So with the simple model, you can run not only three or five scenarios but thousands.

You're also working on regional climate services for Switzerland as well as on extreme weather events in general. Is that where you see the future of climate science?
I think there will always be people who work on the fundamentals. Nevertheless, the more accurate climate models become, the more likely they are to have utility in societal decisions about adaptation to climate change. A farmer in the Limmattal can't do anything with the global temperature, but if you calculate the probability of frost in April and how it will change, that helps him. We have to improve the models to produce these relevant variables.

Besides calculating these climate services, communication is equally tricky. For climate services to have value, it is critical that people see, understand, and use them. It was much the same with weather forecasting: It took us 40 or 50 years from a technical perspective to develop a good weather forecast, but it took us just as long to get people to understand what weather forecasts can and cannot deliver. Today, farmers know that they will make better decisions when they check the weather forecast for rain before making hay. Another example is the decision-making in the case of extreme weather: Decisions are completely synchronized from the fire department to the crisis teams to the health facilities. And although the population is larger than ever before, fewer people die due to natural hazards because we have learned to reduce the risks and predict them better. With climate services, it's a similar story. The calculations are our part, but how the product gets to the customer and how the customer decides to react is an entirely different matter.

Is the research on climate adaptation already being applied today?
That depends on the field. Hydrology, for example, is very quantitative and has calculated its extreme value statistics for a long time, using 1960–2000 as a reference period. However, there is a problem because the system has become a completely different one. Therefore, extreme value statistics are now continuously complemented with the latest data and future scenarios based on extreme value distributions. In contrast, in agriculture and forestry, decision-making processes, such as which crops should be grown where, are still in their infancy. Heat waves have sensitized the health sector, and floods have raised society's awareness for climate adaptation. However, people need to implement climate adaptation measures, and these people must be found first.

Would you say that even more accurate climate models imply not only scientific but also additional societal value?
Again, it depends on the area. We could be much further along in climate mitigation and adaptation based on what we know today. At the same time, that doesn't mean we can't do more with better predictions. We can put photovoltaic panels on rooftops, for example. Nevertheless, it is helpful if someone still thinks about how we can further increase efficiency so that the same panel will produce more electricity in the future for the same price. That's why I believe that there is always a need for people to do basic research, even if the benefits of the results are still unknown.

Today, I work much closer to practice than I used to. Since the translation of theory into practice is urgently needed, I feel I can make more of a difference that way. However, few really feel responsible for that translation. The cities and municipalities have no capacity to derive their actions from research, there is too little money for consulting, and in science, it is not rewarded. As a result, much of the knowledge is left lying around.

We asked you to bring your favorite graphic. Could you describe it briefly?
This cartoon depicts two epidemiologists, with one saying to the other, "As long as we provide the facts to the American people." The climate scientists in the background are rolling on the floor laughing. I use that cartoon in every talk. First, I present the facts: we know the scenarios, we have known the need for net-zero emissions for a very long time, and then I show this cartoon and ask, "So, why doesn't it work?" It's not that climate scientists are more intelligent than virologists. Climate scientists have simply learned over the past 20 or 30 years that political decisions and social actions do not necessarily follow from the facts. People sometimes say "the facts speak for themselves," but the facts just don't speak for themselves. There are the facts, and then there is the question of how to respond to the facts or how to get from

the facts to decisions. That question, along with all my climate research, is what's on my mind the most right now, as well as what role science and the numbers it provides play in the debate.

What would you say science can and should do or not do?
I used to think, like many people, that there was a strict division: science provides facts, and politics decides. That's IPCC's motto: "policy-relevant, never policy-prescriptive"—but that doesn't work. Climate science today cannot be unpolitical at all. For example, if I say, "Switzerland has to reduce CO_2, and it is not doing enough today," then this is a totally scientific statement based on the Paris Agreement, which Switzerland has signed. At the same time, the statement is completely political because it tells politicians that they have to do more. Science of course cannot dictate how the goals should be achieved. This creates tension. The problem here is that in many countries, politics and science don't talk to each other sufficiently and don't cooperate optimally.

However, science has to be part of decision-making processes that also involve representatives from business, lobbies, and associations, as well as environmental protection organizations. If science does not interpret the numbers in the political process, then the numbers will be interpreted by others. Since science is more neutral, objective and impartial than anyone else, it is allowed to make these interpretations. But in doing so, we also have to tell a story and explain what the numbers mean. This explanation is always a little subjective.

Has climate science stayed out of the discourse too much over the last 30 years?
It has certainly held back and taken a relatively modest role for a long time. The IPCC has been one of those examples where governments have said, "We want scientific reports, but please never be policy-prescriptive." So you never find in those 1,000 pages of IPCC reports a statement of what is better or worse; you can't even color the worst-case scenario in red. Now young scientists have internalized this motto of "never be policy-prescriptive".

At the IPCC conference in Stockholm in 2013, you helped negotiate that the CO_2 budget remains in the report. Can you tell us your impressions of the conference? How much is actually being cut out?
It's very rare that things are cut out. I can only remember two cases. First, in Working Group 3 of the IPCC, which deals with mitigation options, the distinction between industrialized nations, developing countries, and emerging economies was taken out because it was felt that this would not serve the political debate.

Second, in Working Group 1, which deals with physical foundations, one of the few words that was crossed out was "pre-industrial" in the fifth report. Since pre- and post-industrial were somewhat arbitrarily defined with the invention of the steam engine in 1750, our attempt to define the temperature rise between 1850 and 1900 as "pre-industrial" was not successful. This avoided setting the period between 1850 and 1900 as the reference level for most climate targets. A year later it was done by the way, so politics is sometimes unpredictable.

We defended the figure of the CO_2 residual budget from 10 PM until 4 AM during the last night of the acceptance plenary. It shows temperature increase as a function of cumulative CO_2 emissions, allowing us to infer a CO_2 residual budget depending on the temperature target. This was too political for the countries of China, Saudi Arabia, and Brazil. Interestingly, temperature increase and CO_2 emission trends were not a problem in two separate plots. On the other hand, other actors like the EU wanted the graph of the CO_2 residual budget to come in. Everything else was already known. But this realization of how CO_2 and temperature depend on each other and that a net-zero for emissions follows from that was new. That’s why, in the end, it was more of a tug-of-war between the countries.

How did you manage to convince the respective countries?
The question is what the governments can do against you. On the one hand, they can try to criticize the graph as incorrect and argue that there is a weakness in your analyses; that is, they criticize that you have not done a clean enough job scientifically. So we spent hours justifying what was done and why. Unsurprisingly, the report has pages of explanations because the pushback was expected.

On the other hand, they may denounce your calculations as incomplete. Sometimes you also hear “It’s not relevant to my government,” “I can’t explain that to my minister,” or “It has no practical significance.” Then you just have to stay technical and friendly and keep trying to explain it.

What is your opinion on peaceful climate activism and civil disobedience?
I have read and thought a lot about it, even though it’s not my research, and I don’t think there is a clear answer. The actions, of course, generate a lot of attention and shake things up. When you call for something extreme, less extreme things suddenly seem quite okay; the spectrum of options shifts. In that sense, I think climate activism has its justification and impact.

On the other hand, climate activism polarizes and divides society even more. Those who are already in favor of it will simply be even more in favor of it. I know many people, for example, my relatives, who took part in “Fridays for the Future”—it was joyful, it was positive, and it was inclusive. But since there have been more radical climate movements, they say, “I no longer want to be linked to them.” Due to this radicalization, climate activism is losing popularity.

I personally don’t participate in these things. If other people glue themselves on the street to protest, then they can go ahead. As a scientist, however, I see this as problematic. I can’t be neutral and objective, talk to people from all parties, and go out on the streets at the same time.

In an interview, you once quoted Daniel Kahneman’s phrase “No one ever made a decision because of a number.” What story would you tell a climate denier?
Facts do not speak for themselves; they always have to be seen in a social, political, and economic context. How one decides based on the facts is a question of priorities, values, personal convictions, and the opinions of people close to us. We buy a Tesla because the neighbors also have one. People rarely make decisions based on numbers but instead consider what they’ve seen or read and thought is cool. What a story that

convinces someone looks like is not so simple. There can be a lot of different stories told by different people. You might be attracted to the Tesla story; someone else might be attracted to the photovoltaic high-tech story. Another person, for example, from the *Last Generation*, might be persuaded by an issue of equality and fairness.

But the question was what story to tell a climate skeptic. Forget it. Don't waste your time trying to convince people who don't believe in climate change for ideological reasons. You won't succeed. That's a collision of worldviews and identities.

How do you deal with the hostility you occasionally experience? Are you sometimes afraid of being insulted on the street?
I read the hate messages and then put them on the pile over there *(Knutti goes to a shelf in his office and pulls out a thick stack of letters that he has collected over the years. The letters are handwritten, typed, or consist of a collage of pictures)*. Most people would never say what they write to my face.

Being actively approached on the street is something you have to get used to. It happens to me about once a week—in the grocery store, on a hike in the mountains, in the train. Most people who recognize me are interested in the subject and have probably read a lot or seen my picture somewhere. Some then approach me and really want to know more. I have rarely had negative experiences.

However, I also have the feeling that the debate has changed. Five to ten years ago, climate researchers were still considered left-wing treehuggers. Climate denial was insanely prominent. Today, it's different. Those who don't believe that climate change is a problem are a minority that is dying out. The discourse is now much more about the rules and costs, that is, how we can achieve the net-zero goal. The fact that we need to achieve the net-zero goal is clear to virtually everyone. With "Fit for 55," for example, the EU has ambitious goals.

What could you see yourself working on in ten years?
There are many different things. Advances in technology will make things in climate modeling possible that we can't do today, especially in basic research. In addition, we will develop climate services much further, similar to weather forecasting, which everyone knows exists today and from which there is a direct benefit. The topics will also evolve in a more interdisciplinary way. For example, today's energy experts still don't consider what exact climate extremes will be possible in 20 years.

Personally, I will probably continue to work at the interface between science, politics, and the public. It is clear that as a university, we educate people and lay the foundations in research. Nevertheless, I think the question of the role of science, universities, and the scientist as a person in the social landscape is absolutely central. Science must do more for the crucial issues of our society in order to master these challenges concerning health, digitalization, the whole environment, and sustainability issues.

When you look at research and societal development, does anything also make you feel euphoric?
I don't know if I can be euphoric. During the last 30 years, everything seemed to be improving. There was economic growth, democracies were created, and everyone

was doing wonderfully. At first glance, everything was getting better. However, the last few years have shown that the future is not simply an extension of the past but is sometimes completely disruptive. Crises are not in our imagination; they are real, and we are poorly prepared for them. For almost everything. And it's not that people didn't know. It's a lack of willingness to deal with complex issues and information.

There have been many efforts by policymakers to better prepare for crises, work with science, and plan ahead so as not to walk blindly into the next disaster. The warnings in recent years have been sufficient. But whether we will really succeed, I don't know. Crises are also quickly forgotten. But I hope that we have learned a little from the last few years and are taking a more systematic approach to climate change. The most dangerous thing would be to wait and do nothing.

Besides climate research, I'm trying to find people to help shape this process. It won't be an easy road, and it's unclear who will be responsible for it and pay for it. It's an ongoing but exciting process because you come into contact with politics, business, and other stakeholders. It's a bit like playing chess with rules that change.

What advice would you give young people interested in fighting climate change?
Try to make a contribution with your personal behavior. Of course, this doesn't just apply to young people. I think everyone can probably reduce 20–30% of their CO_2 footprint in their daily lives by being a little less foolish, driving a little less, eating a little less of animal products, or flying a little less.

At the same time, I am convinced that we cannot solve the problem through individual responsibility alone. We have never solved such a problem in the past with individual responsibility or spontaneous technical innovations. Ever. From waste to sewage, air quality, the hole in the ozone layer, all the way to the COVID-19 pandemic, these problems have always been solved by regulations. You have to build the whole social system in a way that makes it easy and attractive for people to change. Or you have to dictate it to them.

You can advocate for that policy framework. Therefore, be active! Anyone can get involved. If a lot of people help to shape it, then it will probably be better than if everyone just acts alone. So get involved where you think it's important.

Chapter 2
Ruth Cerezo-Mota: "Recognise Your Own Biases"

Johanna Kinder, Ulrike Richter, and Karolin Stiller

©Ruth Cerezo-Mota

Ruth Cerezo-Mota is a climate scientist specializing in regional climate models with a background in oceanography. She wrote her Ph.D. about the mechanisms controlling precipitation in the North American Monsoon in 2009 at the University of Oxford. Since 2014, she has been a researcher at the Universidad Nacional Autónoma de México (UNAM). Furthermore, she was a lead author of Chapter 8 on changes in the water cycle in the sixth assessment report (AR6) of the Intergovernmental Panel on Climate Change (IPCC) Working Group 1.

Date of the Interview: April 25, 2023

G. Lohmann, *Conversations on Climate: The People Behind the Science*, SpringerBriefs in Climate Studies, https://doi.org/10.1007/978-3-032-19147-2_2

What sparked your interest in climate science and oceanography?
What I remember is that when I was around 13 years old and only knew Mexico City, my best friend's brother went to study oceanography, which was only offered in Ensenada. Once, when he came back at the end of the semester, he brought a book with a seaweed collection. I was so fascinated by the variety of seaweed, ranging from green to brilliant, gorgeous pinkish colors and from soft to tough. The university is just on the shoreline, so you can actually go to the beach in the breaks between classes. So wonderful! So, I went and did my bachelor's in oceanography. However, the supervisor of my master's thesis, Tereza Cavazos, worked with climate models. So even though I graduated in physical oceanography, in my thesis I didn't work with the ocean at all. Since then, I've been working on regional climate models and extremes regarding either precipitation, droughts, or temperature. I really don't know why I wanted to be an oceanographer, but it sounded so wonderful and magical—and it actually was; I have no regrets, and I don't think I would have been as happy studying anything else. It all started with "Oh, I want to touch the seaweeds."

Have there been any turning points in your life and career?
When I started my master's, I wanted to do something with the ocean. My thesis for my bachelor's was related to El Niño, and I wanted to pursue that kind of research. But then I met Tereza Cavazos. She was just coming back from a postdoctoral position, and I was her first student. For me, this was a key point in my story because she was so supportive and always had time for me. She was my academic mom, mentoring me beyond just work-related things on personal stuff as well. Even today, whenever I have some challenge in my work, she is the first person I speak to. We became really close friends. Actually, I'm going to visit her for my next holiday. I don't know if she was expecting that when she accepted me as a student. So I'm very grateful for that, I think she encouraged me to do many things I wouldn't have done without her. The second important point in my story is related to the IPCC. There were 20 Mexicans taking part in the whole sixth assessment report, but I'm the only one from Working Group 1. And I'm not just the only Mexican, I'm the only one from Central America. Because of that, when the report was released in 2021, I got a lot of attention and was invited to give interviews and speak about the results to several different audiences. Actually, before meeting with you, I had a presentation with Scientist Rebellion where I was allowed to speak about climate change through different scenarios.

Did you have any personal doubts regarding your work?
When I was young, I was really good at memorizing stuff, so I was always confident. I might not be cute or beautiful or anything, but I was the clever one—until I finished my bachelor's thesis. The committee was made up of some very awful macho men who treated me horribly. At some point, one of them told me that my thesis was terrible and an insult to every oceanographer in the whole world. How would I dare give them that material to read? Even if I were dumb, I should at least try to do my best. That destroyed me completely; I lost all my confidence. Whenever I get an offer, for example, to be part of the IPCC, I still hear the voice of that man in

my head telling me, "You're not good enough. You're going to fail." It happened 20 years ago, and I'm still fighting his voice. It broke me.

What an asshole.
I know. It was really frustrating because when he said all those things, I was speechless. I was not even able to confront him and tell him I'm not dumb. Instead, I started to cry in front of him. So that was an awful moment. When I finished my Ph.D. in Oxford, I wanted to go to his office and show him. Unfortunately, he died before that, so I couldn't *(laughs)*.

How did you regain your confidence? What gave you the power?
At the same time that one person broke everything inside me, I found Tereza, and she supported me no matter what. If you find someone who believes in you, that's enough to be able to focus on the good things. So at the same time I hear his voice in my head, it gets drowned out by Tereza's. Moreover, my mom and my sister have always been there for me; they've been my rock. Whenever I hesitate, they tell me that I can do it and that "We are with you and we believe in you." So I think if you can find some support, even if it's only a little bit, that's enough. And you should try not to listen to the dark side. You are good enough.

What do you think would be necessary to establish a feminist (climate) science?
I think we need representation. Science in general is a very male-dominated world, especially physical science. When I participated in the IPCC, there were multiple women in leadership positions, like Valérie Masson-Delmotte and Carolina Vera as co-chairs, as well as some others like Ko Barrett and Thelma Krug. They lead by example, not in a condescending way. They listened to everyone and gave importance to your voice. This was an atmosphere of respect and equality. For me, it was very important to see them show how we should behave as women in science. I really hope that I can follow their example; they are my role models.

What do you think would be necessary changes in the scientific world to enable more women to conduct research?
I think combatting gender bias in the scientific world has to start in schools and at home. There shouldn't be any differentiation between what boys and girls can do. Regardless of your gender, it should be you who chooses to play with dolls or with cars; both are fine. Also, that sort of "women cannot do this and that" has to disappear. Because we can do whatever we want. I mean, we are not better, but we definitely can. And I think representation matters in all senses—color, gender, and how one identifies. It matters because you realize "Okay, if she can do it, I can do it. Right?" Hopefully we will get more women into science, and then it will be just normal. No one would have to ask why I'm here; gender wouldn't matter. You are a researcher, regardless of the gender you identify with.

What powerful words.

You've already mentioned several parts of your research. Are there specific results of your research of which you are particularly proud?
Well, it's actually not *my* research that I'm particularly proud of, to be honest. I'm currently supervising a young woman, Marta Rodríguez-Gonzáilez, for her Ph.D. on extreme events of rainfall in Yucatan. But for her Master's thesis Marta worked with heat waves and a tool called self-organizing maps, some sort of machine learning. For many years, I wanted to work with that. But if you're a researcher, you have so many things to do and there's very little space for you to learn new things. That's why I proposed to her that she work with this tool. I didn't know how to use it, but she said, "Okay, I'll do it," and she learned how. In her thesis, she analyzes the atmospheric patterns that favor heat waves here in Yucatan. In general, it is very hot here, but when we have a heat wave it's even worse. It feels like you are in hell. Her work can help the meteorological office predict heat waves well in advance. Hopefully our manuscript will be published soon because her results are very good. The idea is that we or someone else can use it, also for other climatic phenomena and other Mexican regions.

As a researcher in regional climate models, how do you define climate regions? Why is it important to define those smaller regions in order to understand the whole climate?
Exactly. Regional climate models are just defined for a small region. You use them for two main reasons: first, because they are of high resolution. Until ten years ago, global models had a resolution of around 50 km, missing small-scale environmental processes. However, if you use a regional model, you can go to really high resolution, such as using grids of 1 km mesh size. Thus, the model will be able to explicitly solve finer and more local processes. Secondly, there are very few institutions that have global models because it's very complex to build one and requires a lot of computational power. We, the people of the Global South, may not have this power. Therefore, regional models offer a solution for how to run simulations on your own laptop, so you can do experiments for your own region with less computational power. This is also the main reason why I started working with regional climate models during my master's. Back then, only around three regional climate models existed; I worked with MM5. I really liked being able to zoom in on things that global models don't account for. Since then, I have been working with regional climate models, mostly for Mexico. For instance, my Ph.D. was about the monsoon rains in western Mexico, and when I moved here to Yucatan, I started to work with the local rainfall and heat waves. That's the beauty of regional climate models. You can easily implement the same methodology in a different area, even if you don't have a lot of resources.

But then how do you include large-scale phenomena such as the monsoon you investigated during your Ph.D. into regional climate models? How do you combine those two different scales?
For a regional climate model, first of all, you define your domain, like Mexico. Next, you need to provide pieces of information that are called boundary conditions and initial conditions. This information about the boundaries of your domain comes from large-scale global models or analyses. Additionally, there is a zone called the buffer zone in which the transition between the different resolutions takes place. Then you can run the model with its iterations and equations. All models, regional and global, solve a set of six or seven equations that are called primitive equations. They are the basic equations that you need to solve to be able to make weather forecasts, climate predictions, and the like. That's how it works: you incorporate large-scale global models into the boundaries of your domain and solve the primitive equations at each time step and each grid cell of your regional model.

Why is rainfall a crucial aspect of regional climate models?
Rainfall is the basis for the existence of life. It provides ground infiltration, groundwater, and water for plants. Before my master's I didn't know it was such a complex process. I just thought you had a cloud, it gets dark, and voilà, it starts to rain. But you need a lot of ingredients to actually get precipitation. The problem with climate change is that the seasonality of rainfall is shifting, if not changing completely, because we are altering everything. Thus, precipitation is happening more in extremes now: Instead of raining the whole rainy season, it rains fewer days, or it rains a lot. When it rains a lot, it exceeds the capacity of the ground to infiltrate water. Therefore, the aquifers do not get recharged as they used to. Instead, the overland flow increases with flooding and more evaporation as consequences. Global warming is changing precipitation patterns and, thus, all the other processes and components of the global water cycle. That's why it's important to understand how it happens, why it's changing, how it will keep changing, and its resulting impacts. When we are talking about food security, we are actually talking about water security because we need water for irrigation. Rainfall is related to absolutely everything.

We asked you for your favorite figure, which is also about water cycle changes. Why did you choose this figure?
The figure is part of the FAQ of the 8th chapter in the IPCC Sixth Assessment Report (AR6). I wrote the answer together with my friend and colleague Paola Arias from Colombia. I'm very proud of our work because it took a lot of time to answer such a complex question in simple but not simplistic language, with only one page of space. After we had written a first draft, someone working in science communication tried to render it more appropriately and then it went back and forth until we were happy.

At the same time, we had to create a sketch for the figure accompanying our text. We used many arrows going everywhere in order to convey a lot of information. It was quite chaotic, but then a designer with a scientific background created this wonderful final figure out of our sketch. I'm very happy with it because it is not this typical sketch of the water cycle with rainfall that leads to rivers flowing from the mountains to the oceans, evaporation leading to rainfall, and so on. Instead, we needed a water

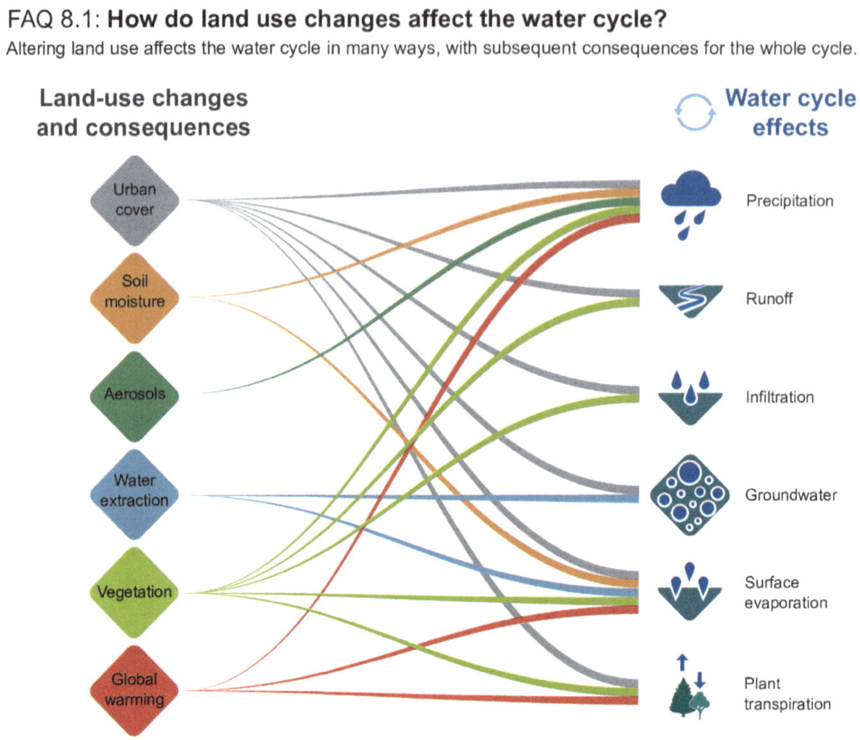

Fig. 2.1 Land-use changes and their consequences on the water cycle. As all the components of the water cycle are tightly connected, changes in one aspect of the cycle affects almost all the cycle. Figure from IPCC AR6: FAQ 8.1 (Douville et al. 2021)

cycle that included all the processes and complex direct and indirect relationships affecting it. We struggled to convey this idea, but the final figure illustrates perfectly how everything is related: if you modify any of the components at any of the scales, you will affect everything else—a balance cycle (Fig. 2.1).

What are the most important aspects that affect and change water availability on a local scale?

Besides everybody's responsibility to reduce emissions, the conservation of land has a direct impact on rainfall. For example, if you deforest some region, the fact that you remove the vegetation is going to affect some process that happens at a very fine scale. You need vegetation to sustain the rainfall season. If you removed all of the vegetation here on the Yucatan Peninsula, you would still have precipitation because of the tropical cyclones, but you couldn't sustain the rainy season. There are several feedback mechanisms between vegetation, the ground, and the water that is stored in the first meter of the ground. Even if you substitute the removed vegetation with another type of vegetation, those mechanisms will definitely be altered, not to mention substituting it with asphalt or something similar that prevents infiltration completely.

What is the biggest knowledge gap right now for modeling precipitation in regional climate models?

As I said earlier, all models solve a set of equations, but we still have different results for each model. One reason is differences in the numerical schemes. Another reason originates in something called *parameterizations*, which is a way to implicitly describe some environmental processes without using equations in your model. With this, you can handle the complexity of all the various interactions between the involved processes in regional and global models. However, the big problem is that these parameterizations were created in places like Europe or the United States. Even though the physics is basically the same as in the high latitudes or tropical regions, the processes and therefore the parametrizations are not.

Yucatan, for instance, is very peculiar because during the summer we have the influence of trade winds coming from the Sahara and bringing dust aerosols. During winter, we are influenced by cold fronts from the States that carry marine aerosols, like salts, with different characteristics than dust. These aerosols are necessary for rainfall because water vapor requires the surface of a particle to condensate, but the process depends a lot on the size and other characteristics of the individual aerosol. By chance, I found out that a colleague of mine, Dr. Luis Ladino, collected data on these aerosols here. Right now, another Ph.D. student, Salvador Castillo LIñan, is trying to use Luis' data to improve the parameterization of our regional model. But the fact that we are working with data specifically for Yucatan is precisely why our model won't necessarily work in another tropical region. So, we will always need local research done by local people.

What is your overall goal for your research? What do you want to achieve?

I hope that we can provide better numerical schemes, at least for some specific areas, so that the operational centers in Mexico can provide better outlooks and projections on a seasonal scale. We want to reduce uncertainty. Hopefully, policymakers can then make more informed decisions. For example, when it comes to why we have to preserve a certain green area, such as a forest or the jungle, there has to be information on its importance for our survival—it's not just green and nice to have. I do hope that we produce results that help to avoid the most catastrophic scenario for climate change here in the region. Normally, I try to present the result of every research project to policymakers at the end so that they at least know that some information exists and that they can contact me. That's the final goal: to increase knowledge and make informed decisions.

We saw that you shared political posts on your Twitter account. What does and doesn't work well right now in climate politics in Mexico?

After our last change of government, the new, allegedly left-leaning president, Lòpez Obrador, turned out to be actually very right. Although he had never talked about climate change in any of the debates of his campaign, he had promised more funding for science, education, and health. But since he became president, he has given a lot of power and control to the military. Instead of increasing funding for science, technology, and climate-related disasters like hurricanes, it has been cut, and there is no information about where the money is ending up now. So there is no transparent accountability and a lot of corruption.

Is climate change a topic that is discussed a lot in public?

As a response to the government's strategies, a lot of movements and collectives have emerged that create open spaces to talk about climate change and its effects. I think, for most people, talking about climate change means talking about polar bears: "Oh yeah, it's very sad, they're going to die, but they are far, far away. We don't care." But that's not the case; it's not even in the future. It's now. We are already facing the impacts, like more extreme weather events, and it will get worse if we don't do something now. But there are also a lot of people trying to share this idea.

Last week, I got invited as part of a group to present the government with a 40-page handbook. This handbook contains information from the IPCC reports and was translated into Spanish. It is very simplified and very focused on Mexico. It also includes strategies on how to act if you see something that shouldn't be there—for example, when you notice that some region is being deforested or polluted and you want to file a lawsuit. In the book, you'll find instructions on how to make a complaint to the authorities so that they will be forced to at least go to the site and investigate if there is something unlawful happening. The document is very nicely designed with a lot of graphics because we want all people to have access to the information, not just academics. We are all part of the problem of climate change, not just policymakers. Thus, we all have to be part of the solution, and the only way to be part of the solution is to have the knowledge and know-how to do something.

How can we use our knowledge about climate change to actually implement change in society?

I'm very lucky to have a lot of opportunities to talk about climate change. On the one hand, I provide information that shows that we need to act, adopt a diet with less meat consumption and less food waste, and demand that governments attend to their duties. However, although we have to acknowledge that we are bad at solving the problem, I always try to close with a positive remark that there is hope.

When we were finishing the report and locked up in our homes due to COVID-19 pandemic, I got depressed, couldn't work, and was about to quit my job. I was not able to open my computer or answer emails because I felt dizzy and sick. At some point, I had 666 emails to read. But I couldn't do anything about it because those feelings crippled me. When three of my colleagues and friends from my chapter in the IPCC contacted me separately, asking if I was okay and alive, I felt really bad about their being worried. That sort of brought me up, and I started to get reincorporated into my work again. Later, one of those three colleagues told me that she had felt similarly, but her motivation was to make sure that people who have to make the decisions will not have the excuse that they didn't know. This is our contribution, as small as it can be. This is how we are fighting climate change. So I promised myself that whoever invites me to talk about climate change, I will do it, because that's the thing that I can do.

Is there a result of your work that has particularly contributed to the fight for climate and nature?
My students and I do research on the relevance of greenery and the need to preserve vegetation to sustain rainfall and prevent unlimited temperature rise here in the Yucatan Peninsula. The government is removing a lot of vegetation to build a new train route, the Tren Maya, but this will fracture the ecosystem, as some species will not be able to cross from one side of the rail to the other. When they started this project, they promised not to remove one single tree, which we knew was impossible. By now, two years after they started, the official number of uprooted trees has reached three million. Activists and collectives that have been monitoring this project talk about ten million uprooted trees. This will immediately lead to more heat. So the contribution of my research is to show that projects such as the Tren Maya are not going to solve problems, but they are going to create many problems in a really short time.

Why does the Mexican government want to realize this project?
To give a bit of background, all the permits to build the train, the railroad, and everything else were directly given to a private company called Vidanta without any public bidding. This company gave a lot of money to the actual president during his campaign. We don't even know how much money they are receiving for this project now. The government is also giving a lot of permits to the military. Actually, the military is building a hotel in the middle of a reserve that will be administered by them. Thus, the president is giving a favor to Vidanta and to the military so they will protect him in case something happens when we change government. At least I don't see any other reason for what he is doing.

How do the people of Mexico show their protest?
The problem is not the train path itself, but they began building it without having done any proper studies about its feasibility. The Yucatan Peninsula is karstic, so we have a particularly calcareous and porous sort of soil. It does not have the capacity to support the weight of a train, and it has already collapsed on some parts of the path. Many activists and collectives here on the peninsula invite people to go with them to a particularly problematic passage in the jungle near Tulum. There, people can see the deforestation and how the aquifer is being polluted by being filled with concrete. Many well-known people are advocating as well. There are also demonstrations in Mexico City and a lot of activity on Twitter to show what is happening. It is important to preserve this region and the reservoirs in order to live on this peninsula. Otherwise, there will be consequences for the whole country, such as migration.

How did you get involved in activism?
First of all, I did not expect to become a speaker about climate change. Once, the United Nations Against Organized Crime invited me to a panel about illegal trafficking to provide a physical explanation of climate change because its effect on the wildlife situation is exacerbated by wildlife trafficking. Among the other speakers was a young climate activist named Aurelien. He then invited me to several other panels and included me in a WhatsApp chat with other activists. I definitely don't

consider myself an activist, but I'm very happy to be there because I relate to what they do, and I am happy to help in any way I can.

How do you think the international community can support those frontline activists in Mexico?
The bad thing and the good thing about our current president is that, like Bolsonaro and Trump, they want to be popular—the loved one, the chosen one, the leader. And when demonstrations, such as those against the Tren Maya, are replicated in countries such as the UK and Germany, he will feel urged to at least pretend to do something. That would already be a tiny little victory. In this way, international support can put pressure on the government.

Twitter and other social networks are also a good way for activists to connect, exchange information, and meet or follow people they relate to. On the other hand, it is also used by climate deniers to disseminate information, bully people like high-profile climate scientists, and share really evil things. Misinformation and fake news seem to get replicated much easier than well-informed news. You just have to be careful, try to get a trustworthy source of information, and verify the information as much as you can.

We already talked about the political dimension of science. In what way does post-colonialism create injustices in the global scientific community?
One thing is collaborations with people from the Global North coming to Mexico, Latin America, or Africa. They use us to get the knowledge they want, but then go back to their countries, publish, and don't give us much acknowledgement, often not even including us as authors. Once, I had a project funded by the British Council with the condition that it include a principal researcher from Mexico and one from the United Kingdom. Even though I was the one writing the proposal, the project appeared in the CV of the British scientist without my name or our collaboration anywhere. No recollection, no recognition, nothing. There are worse examples, like the concealment of authorship or the treatment of people as subordinates. That probably happens not on purpose but through unconscious biases we all have. We need to be aware of those biases, work on them, and treat each other as equals.

Another thing is the way the IPCC works. Literature that is not written in English will not be assessed for the report, no matter how good it is. Even for people like me who learned English from primary school on, there might be a language barrier. This comes on top of a systematic bias: in order to be reviewed, you need to be able to afford publishing a paper, which costs, for example, around $10,000 in *Nature*. Of course, you cannot include anything that is not peer-reviewed. There has to be a way to incorporate non-English research into the assessment because we have to take local people seriously. As an example, we wanted to create a specific graphic for the North American Monsoon that shows how much the precipitation regimes change per region. In the first draft of our figure, the extension of the monsoon covered all of Mexico and Central America, which is wrong. So I provided scientific papers, and without any issues, the figure got adjusted. But someone from the United Kingdom who specialized in the Indian and global monsoons interfered by email, stating that I was wrong because the region I proposed was too small. So we started to exchange emails with several people. There I referred to my work on the monsoon during my Ph.D. and tried to convince him, even though I had only five papers published and he had 100. He said that our opinion would be misled because instead of doing our own research, we would use US data that didn't focus on Mexico. That really made me angry. Thankfully, one colleague from my chapter backed me up and sent him 50 papers on the North American Monsoon for evidence. Another colleague from Colombia asked him to please focus on the science, which means to get informed before having an opinion. Those sorts of things happen in academia. As a person from the Global South, if you don't have enough papers, your research doesn't count.

What could be done to support researchers from the Global South?
Academia can be a very tough environment, but I think, just like everybody else, you have to respect your own biases. As there were some issues in my working group of the IPCC, the co-chair hired a company specializing in the reduction of biases, like racism, female-male bias, this established superego against someone young, and so on. They gave a lot of talks and workshops and had a meeting in person with each of us. And, of course, I also have some unconscious biases myself—everyone has them. But this is what we first have to acknowledge in order to get the tools to fix it and become a better person. If you achieve that, you can be a role model for others, making them aware of their misbehavior.

You mentioned that you were the only Central American in the first working group of the IPCC. Which were your biggest challenges on a personal level?
I think it's a big responsibility for everybody, but being the only Central American, I felt extra pressure to make it good, although nobody told me to. Whenever there was an issue regarding my region, like a figure missing some information, I felt the responsibility to attend to it. If there had been another person from here, it would have been less of a burden, because I can't know everything, not even about Mexico. To give a quick example: the regional facts sheet is one of the outreach documents created at the end of each report. It has a summary for North America, Central America, Europe, Asia, Africa, and so on. But we were working on it for quite some time in an even smaller subdivision of the regions, so it wasn't until the final document that I noticed Mexico was split diagonally into two. The north and west belonged to North America, and the southeast to Central America. But Mexico had to be represented as a whole because otherwise policymakers would have to download the material for two regions to see one country. It was a really difficult process to achieve this as many authors from the States didn't want Mexico to belong to their already quite large region. In the end, after many emails where I thankfully received some support, the IPCC had to intervene and decide that Mexico should be represented as a whole, either as part of North America—whether the US likes it or not—or as a region on its own. However, the fact that I myself didn't notice it before made me feel terrible. I think it was part of an unconscious bias. I have to remind myself that I don't have to take on all decisions myself. I could have asked my colleagues in Central America.

What do you think are the weaknesses in the way the IPCC is organized?
I think there are two weaknesses. The first is the definition of the IPCC as a "neutral" institution. The IPCC is very political and democratic. All the decisions have to be made in agreement, which is good because then you have to listen to everybody. However, sometimes you have to make a decision, and it cannot be neutral anymore because neutral is a position, whether you like it or not. You are deciding not to do anything. But we have to choose one side—the side of science. We have to say something when we are at the Conferences of the Parties (COP) to tell them that we cannot exploit more oil banks and stuff like that. At the end of the day, we are not the best, but we are the most informed on several topics because we have read so much. Furthermore, we have the platform to say something. So we cannot be neutral anymore in the face of this crisis that we are living in.

The other thing that needs to change is the division into three working groups. At an earlier stage of the IPCC, it made sense, but now it all has to be connected. I think the next report needs to be just one report—not one of 3,000 pages, but one that integrates the three groups. Because whatever affects the climate is going to affect humans and biodiversity. We are all connected. That's why we need to find a way to all work together, even though it may be difficult. We have to break the barriers between the different sciences and find a way to respect all voices. At the end of the day, a whole integrated report is going to be a more useful document for policymakers than three separate ones.

How should the IPCC change the way it selects its authors?
The process of author selection begins with the IPCC making an open call for everyone. This call is then replicated and announced by the Government Focal Points in each country, which give it as much publicity as they can. I, for example, got into the IPCC because, by chance, I knew people from the IPCC who contacted me and asked me to apply. Nevertheless, I was sure that many people would apply and I wouldn't be chosen because I was young. I just fulfilled my duty by applying so I could have the right to complain. But in the end, I was the only one from Mexico. However, I know some colleagues who sent their applications to the Focal Point in Costa Rica, but this Focal Point decided not to send them to the central office because they wouldn't fulfill the profile needed for the IPCC. So nobody from Costa Rica was invited. I think the same thing happened in Guatemala. I think that's a big issue. The fact that you have to go through the Focal Point is the first obstacle because it might be personal. Sometimes the decision is not whether you have or do not have the credentials or experience to be part of the IPCC, but whether the people in the Focal Point like you or not. If the IPCC wants to assure more diversity and more applicants from underrepresented regions, they have to change the mechanism and allow, for example, direct self-nominations.

Which developments in climate research are particularly promising, in your opinion?
Well, I expect that we can provide more accurate simulations for different regions around the globe, as well as for the Yucatan Peninsula. If we manage to do it for the present, the chances of more accurate projections will also be higher, and we

can reduce uncertainties in the scenarios. But that goes hand in hand with having better observations, which means more funding. Field trips and that sort of thing are very important to get observation and validation data, but they are so expensive. You need equipment. You need people. You need money to travel. There are still so many processes that we don't fully understand.

Is there any advice you would like to give people who are just getting into science? I think that you have to go through the world with an open mind—especially in science, because sometimes people are like, "Oh, I know about physics and thus I'm smarter than the people who work in other areas." That's not true. Each area of knowledge is so complex and demands so many skills that they just may not be comparable. Maybe I'm very good at programming, but if you put me in the middle of the jungle to quantify animals, I'm going to be dead in ten seconds. Or if you ask me about law or economics, I don't have any idea.

However, learning about other areas can help you reshape your research and use your funding in a better way so that it actually helps society. Research is not only a philosophical question of "Oh, I want to know something." We want to know something because it has an impact on the real lives of people, and if we understand it, we might be able to change things. I'm not saying that basic science doesn't have any value or application. Of course it has. But when you are willing to recognize that each area of knowledge has its own importance and everything is connected, you can produce a more valuable product in the end. Usually there's not only one thing that is going to solve a problem because each problem has so many layers, and the only way to solve it is by checking all of them. This involves many processes and a lot of knowledge. So you have to recognize the value of everyone and try to listen and learn as much as you can from the colleague next to you.

Reference

Douville H, Raghavan K, Renwick J, Allan RP, Arias PA, Barlow M, Cerezo-Mota R, Cherchi A, Gan TY, Gergis J, Jiang D, Khan A, Pokam Mba W, Rosenfeld D, Tierney J, Zolina O (2021) FAQ 8.1 Figure 1 in IPCC, 2021: chapter 8. In: Masson-Delmotte V, Zhai P, Pirani A, Connors SL, Pean C, Berger S, Caud N, Chen Y, Goldfarb L, Gomis MI, Huang M, Leitzell K, Lonnoy E, Matthews JBR, Maycock TK, Waterfield T, Yelekci O, Yu R, Zhou B (eds) Climate change 2021: the physical science basis. Contribution of Working Group I to the sixth assessment report of the Intergovernmental Panel on Climate Change. Water cycle changes. Cambridge University Press, Cambridge and New York, NY, pp 1055–1210. https://doi.org/10.1017/9781009157896.010

Chapter 3
Stefan Rahmstorf—"As Scientists, We Have a Duty to Society"

Julius Mex, Johanna Kinder, Leon Focks, and Alexa Beaucamp

©Astrid Eckert

Stefan Rahmstorf is an oceanographer and climatologist at the Potsdam Institute for Climate Impact Research (PIK). Since 2000, he has also taught "Physics of the Oceans" at the University of Potsdam. He studied physics in Konstanz and Ulm, Germany, and received his PhD in oceanography in 1990 from the Victoria University of Wellington, New Zealand. After working for the New Zealand Oceanographic Institute and the Institute for Marine Sciences in Kiel, he started working at the Potsdam Institute for Climate Impact Research in 1996. His joint blog *realclimate.org* with international colleagues was described as one of the top five science blogs by

Date of the Interview: March 27, 2023

G. Lohmann, *Conversations on Climate: The People Behind the Science*, SpringerBriefs in Climate Studies, https://doi.org/10.1007/978-3-032-19147-2_3

Nature magazine. In addition, he received the Climate Communication Prize of the American Geophysical Union in 2017.

Is there something that is particularly on your mind today?
Yesterday, a referendum on more ambitious climate legislation was held in Berlin. To be honest, I was surprised that so many people went out to vote and that so many rejected the proposal. I had expected that the people who were not on board or not interested would stay at home, so I was confident that the referendum would be accepted when the high turnout showed during the day. But almost half of all voters voted against it.

This is probably due to the fact that there have been many voices in the media talking about costs and telling people that, for example, rents would rise drastically if all houses had to be renovated. That could have been a motivation to go and vote against the referendum.

What was your motivation to become a climate researcher?
I’ll start with my motivation to study physics because that was actually the beginning. I already knew at the age of 12 that I wanted to study physics or at least become a scientist. For whatever reason, I was simply interested in science from an early age and got an experiment kit when I was 10 or 11. I think that was a key experience. I never hesitated, and I was actually convinced throughout my secondary school years that I wanted to be a researcher. Physics, and astrophysics in particular, interested me the most at school, so I borrowed a lot of books on the subject from the university library in Konstanz, where I grew up.

How I ended up in climate research is directly related to the German Academic Scholarship Foundation. As the foundation encourages students to spend a year abroad during their studies, I thought about going to Great Britain after my intermediate diploma. Since Google did not exist back then, I looked at the courses offered by British universities in our university library and came across physical oceanography. I had always loved the sea and was thrilled to see that there was such a branch of physics, so I studied it during my year abroad in Bangor, North Wales. It was there that I decided to pursue this topic professionally later on. I chose my diploma thesis in Konstanz so that it would be compatible with physical oceanography. Of the two options, solid-state physics and cosmology, I chose the latter. I dealt with galaxy formation shortly after the big bang because this is based on the fundamental hydrodynamic equations as the ocean circulation, mainly the Navier–Stokes equation, albeit in a general-relativistic form. For my doctorate, I switched to oceanography and did my doctoral thesis in New Zealand.

What role do the oceans play in the Earth system?
I usually give one-hour lectures on this, but I’ll give a short version for you. One aspect is heat storage on a seasonal scale. Because thermal inertia buffers rapid temperature changes, the climate in coastal regions is usually not as harsh in the winter and not as hot in the summer. A second aspect is heat transport in the ocean. For example, due to the Atlantic meridional overturning circulation (AMOC), the entire northern Atlantic is significantly warmer than it would otherwise be. This ocean circulation

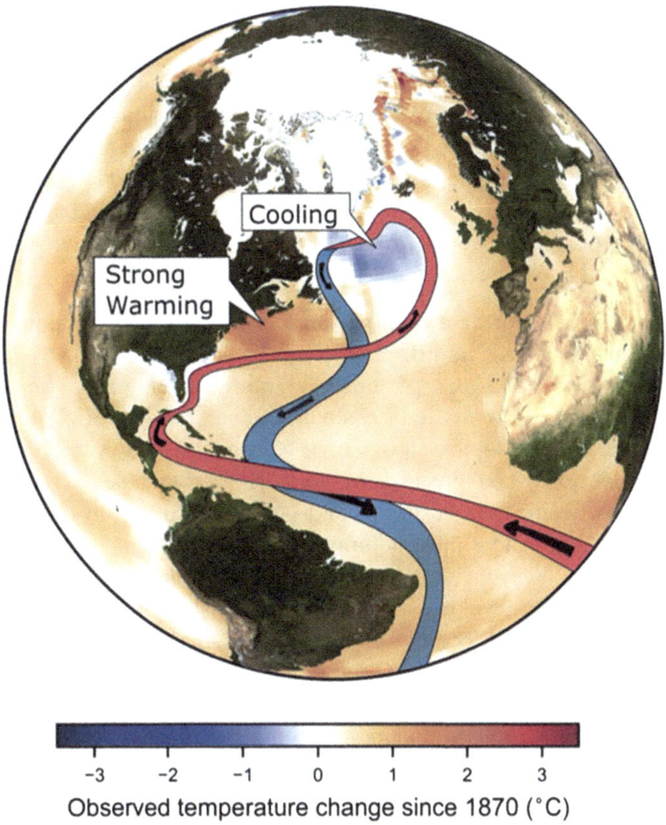

Fig. 3.1 Observed sea surface temperature trend 1870–2016, with a schematic of the Atlantic Meridional Overturning Circulation (AMOC). After Caesar et al. (2018)

also has instabilities and a tipping point. We know this from the history of the Earth. In addition, the ocean is the main source of water for the entire water cycle on Earth. Evaporation from the ocean brings precipitation to the continents. It also plays a major role in other cycles, such as the carbon cycle. The CO_2 concentration in the atmosphere would rise much faster without the ocean because a quarter of what we emit is absorbed into the ocean via gas exchange at the sea surface (Fig. 3.1).

Your research indicates this overturning circulation is weakening. What impact is that going to have on our climate?
The consequences are already noticeable because there is now a cooling region over the Atlantic, or a ‘warming hole’, as it is called in the literature. This is the only region in the world that has cooled down in the last 100 years. This influences the path of the weather systems that come into Europe from the Atlantic. A British study has shown, for example, that the jet stream prefers to go south around this cold zone because there is a low-pressure area above it. The jet stream then arrives from the southwest

and brings heat waves to Europe. This was actually confirmed in the observational data, for example, in the heatwave of 2003.

At first glance, this seems paradoxical because many people think that the weakening of the Atlantic ocean circulation will cause temperatures to fall drastically in Europe and North America, as dramatically portrayed in the movie *The Day After Tomorrow*, but that only applies when it finally breaks down completely. In that case, this cold area, which is now only located over the ocean, would expand to such an extent that it would also cool down land areas in northwestern Europe, especially Great Britain and Scandinavia. This is shown by model simulations.

How did you come to conduct research on so many topics, from paleoclimate to the AMOC, sea level rise, and tipping points?
I started in the early 90 s as a postdoc examining the above-mentioned AMOC. When I started my postdoc at the Institute of Oceanography in Kiel, my supervisor suggested some possible topics. However, he advised me not to work on the AMOC, saying my predecessor had already covered it all—and that's exactly the topic that I took because it seemed very exciting, and I think nonlinear physics is cool. After my postdoc in Kiel, I did almost nothing on that subject for quite a while, but in recent years I have published some papers on it again. In the meantime, I became more involved with sea-level rise and have also published on that subject. I got into sea level research through my work on the fourth IPCC report. There was a lot of discussion then because the IPCC report predicted a fairly low sea level rise of 59 cm by 2100 at worst. The IPCC works based on a consensus principle, which sometimes leads to the lowest common denominator, and in the end a number of colleagues did not agree with this. Among them were David Archer from Chicago, with whom I co-authored a book, along with others I knew and myself, who thought that the IPCC was massively underestimating sea level rise. Then, in 2007, I showed in a paper in *Science* that the correlation between warming and sea level over the last 100 years points to a much higher sea level rise than the model simulations indicated. These model simulations, which were common practice at the time, suggested, for example, that the Antarctic ice mass would increase in a warmer climate due to heavier snowfalls. However, even then, satellite data showed that Antarctica was losing ice mass. This issue brought me to work on sea level research. However, I never had the financial support to continue researching this subject.

I got into extreme weather research in a similar way and studied it more intensively, particularly in the last ten to fifteen years. In the fifth IPCC report, there were debates about the extent to which global warming increases weather extremes. I was unhappy with the approach favored by the IPCC, focusing on attribution studies. This is a methodology that aims to determine how much more likely extreme weather events will become with warming by comparing models. It is thus constrained by what the models can produce, and the ability of climate models to represent small-scale weather extremes was (and to an extent still is) quite limited. While it's clear to us that the general increase in weather extremes is caused by climate change, it is a difficult problem to estimate what role climate change has played in a single extreme event. Often, this methodology finds little evidence, but that doesn't prove that global warming didn't play a role in an event. I am not criticizing formal attribution research per se but rather the fact that the IPCC focused so much on this methodology, which is only one of several lines of evidence. If you take all the lines of evidence together—our understanding of the physics, the observational data from this attribution research, and model simulations—then I argued that the balance of evidence is actually very much in favor of weather extremes already increasing as a result of global warming. In the meantime this is also the conclusion of the latest IPCC reports.

At times, I also did a lot of work on paleoclimate, especially ice age cycles. In addition, I also lead a department here at PIK, which is, of course, connected with many other tasks. Unfortunately, I don't have as much time for research as I would like.

How conclusive are current climate models? Can we trust their predictions?
The first, early models were so-called energy balance models, which exclusively considered the energy balance of the entire planet. This energy balance is quite simple: the energy flux arrives from the sun, and a part of it is reflected right away. The Earth itself emits long-wave thermal radiation, so the result is an energy balance that we change through our CO_2 emissions. So it is relatively easy to calculate how much warming will result. The calculations from the 1970s overestimated this warming, but since the early 1980s, the models have been very accurate. Today, these models are used in combination with many other types of models.

Predicting temperature is much easier than predicting precipitation globally. While large-scale predictions have been reliable for decades, the smaller the scale and the more complex the physical processes, the more difficult the predictions become.

Climate models don't get everything right, but the fundamental trends predicted are rather reliable. One observes that the basic predictions of the climate models have come to pass. That extreme precipitation is increasing worldwide has been predicted by the climate models for 30 years. Likewise, the weakening of the Atlantic ocean circulation has been predicted by the models. The physics behind this is quite complicated, so the uncertainty in the predicted amount of weakening is rather large, but models agree on the fact that such a weakening will happen, and indeed it has already started.

The models also initially underestimated sea level rise, but they now reproduce past trends very well. Likewise, the strengthening of tropical storms and the ice melt in Greenland and Antarctica are occurring as predicted.

What aspects of scientific research do you enjoy the most? Which scientific findings are you particularly proud of?
I particularly enjoy research in the field of paleoclimatology because I find it incredibly exciting that we can find out so much about past climate changes—hundreds of thousands, even millions of years ago. I think it's just amazing what traces have been left in the sediments and in the ice and how you can use them to reconstruct the climate of the past. This is a great and exciting type of detective work. It has also helped us understand that there were drastic climatic changes in the Earth's history, especially the abrupt climate changes during the last ice age which were triggered

by instabilities in the North Atlantic currents. That's how I got interested in paleoclimatology in the first place, because the "paleo community" immediately took notice of my work on the stability of the Atlantic circulation in the 1990s and invited me to paleoclimate conferences.

What are you currently most worried about?
I am most concerned about the effects of climate change on people and our society as a whole, in particular threats to food security and the risk of famines. A second aspect is the irreversibility of our anthropogenic changes, which is not only related to tipping points. Much of the CO_2 emitted until now has remained in the atmosphere for thousands of years. Here at PIK, we do research trying to predict the ice age cycles of the next millions of years. We expect that with our past CO_2 emissions, we have already prevented the next ice age due in 50,000 years. This is to illustrate the enormously long time scale on which we are changing the climate. The new IPCC Synthesis Report and the Summary for Policymakers clearly state that what we do now will determine the climate on this planet for millennia to come. And thirdly, of course, I'm very concerned about the tipping points of the continental ice masses that can trigger a sea level rise of many meters.

Within climate science, where do you see the biggest challenges in the coming years?
I wouldn't say there's one main issue to work on. There are many issues that need to be looked at in more detail. Extreme weather events are a topic that is very exciting. Even though a lot of research has already been done, it is specifically the so-called dynamic effects that interest us at the moment. The thermodynamic effects are quite clear: it's getting warmer, so we expect more heat waves. Additionally, there is more water vapor in the atmosphere when temperatures rise, so there is more extreme rainfall. But then we also expect dynamic changes in weather patterns. How does the jet stream change? What happens to ocean circulation, which then again changes the dynamics of the atmosphere and the ocean? That is still a very open field of research. The other topic, for which we have created a Future Lab at PIK, is tipping points and their interaction—in other words, the domino effect when one tipping point causes another to be reached.

You work on many different topics and also do interdisciplinary research. What is the importance of interdisciplinarity in climate science, and what difficulties, if any, does it pose?
We have a very interdisciplinary approach at PIK. Within my department, we work both in physics and with biosphere modeling. We develop models for the land biosphere as well as the ocean biosphere. That alone can be described as interdisciplinary. At PIK, we don't just do natural science but also economics and social sciences. One of our departments works on solutions to the climate crisis, on systems for socially just CO_2 pricing, on the stability of electricity grids, and on renewable energies. One realizes quickly how important it is to gather the different disciplines on one campus instead of communicating exclusively via virtual research networks. It is essential to meet regularly and exchange ideas across disciplinary boundaries so

that interdisciplinary work becomes a habit. One can also get to know the language used in the other discipline. In different disciplines, certain terms are used and understood differently; they have a different meaning, and misunderstandings can easily occur if people don't maintain regular contact. In this respect, I think it's great that we're all on the same campus here at PIK and that we've been working together across disciplines since the very beginning.

You are very active in the fields of communication and knowledge transfer. How did you get involved in these?
It's very much due to my first year abroad in Wales, which was made possible by the German Academic Scholarship Foundation. There, I got to know the Anglo-Saxon scientific culture. That's also when I started reading the *New Scientist*, for example—a fantastic example of generally understandable science communication at a high level. In the UK, I encountered a great openness to making scientific results understandable for a broad audience. During my PhD in New Zealand, I also experienced that people talk about their scientific results publicly rather than in specialist journals only. There was less of the academic ivory tower attitude. On the other hand, in Germany at that time, it was still frowned upon to talk to journalists as a scientist. Thus, after publishing my first *Nature* paper, I secretly sent a private press release to two journalists to communicate my results. I knew it was frowned upon, but I was influenced by Anglo-Saxon culture. In the 1990s, I published two essays in the *New Scientist:* one pointing out the urgency to stop global warming, and one on the Atlantic ocean circulation tipping point. After that, I was less active in science communication for quite a long time—until the major flooding disaster on the Elbe River, which happened here in Germany in 2002. That was a big turning point for me. It is clear that global warming increases extreme precipitation, hardly anyone denies this nowadays, but it was not self-evident at the time. Articles were then published in the press, misleadingly explaining global warming by changes in solar activity. *Spiegel* even published a corresponding graph that, at that time, had already been retracted by scientists due to a methodological error. I thought that something had to be said about this. Hence, I published an article in the German Newspaper *Die Zeit* after the Elbe flood, and that's how it all began. Later, I got into blogging through my American NASA colleague Gavin Schmidt.

How can knowledge transfer succeed most effectively? What is most important, and what is perhaps still missing today in the communication of scientific findings?
It's completely different from writing for a specialty publication because you talk to a completely different audience. For me, the most important thing when communicating to the general public is that one first has to relate to the audience's state of knowledge. That is what many colleagues find difficult. It is challenging to understand which technical terms need to be explained to a non-expert—they often presuppose knowledge that the non-expert simply doesn't have. That's why you have to put yourself in their shoes. What can you expect from the listener? What do you have to explain, and how can you do that? Explain it as simply as possible without it becoming incorrect. That's what's key in the end.

As the author of the IPCC Fourth Assessment Report in 2007, what was your experience? What role do the Assessment Reports play in the transfer of knowledge? Do you see difficulties in the working methods or structures of the IPCC?
In general, I think it's great that the IPCC exists. For me, working on the Fourth Assessment Report was a positive experience because of the fantastic working group I wrote our chapter with—in my case, it was the chapter on paleoclimatology. It was a great opportunity to network because one works relatively closely with one's colleagues for years and regularly meets, works on the texts, discusses reviewer comments, and so on.

The main downside is that the whole process is extremely time-consuming. The "writing by committee" is very tedious as every sentence is debated.

The process to finalise the Summary for Policymakers is, of course, different because it is a political process. As a regular IPCC author, you are normally not present at this meeting. Each government sends a delegation, which then agrees to the summary. Representatives of the environment ministries are usually sent, but for the Fourth Assessment Report, I was asked by the government to join the process as a scientist.

This approval process for the Summary for Policymakers has advantages and disadvantages. In this consensus process, they really go through the text sentence by sentence, and as soon as one government raises its hand, there is a long discussion about the wording. The lead authors of the chapters of the reports then judge whether a change in wording is covered by the respective chapter. The government delegations cannot, of course, write anything into the report that is not supported by the science. Nevertheless, wording is weakened again and again by governments with specific interests during the consensus process. But once the summary has been adopted, no country can distance itself from it again because they have agreed sentence by sentence, and this is the basis for negotiations under the [United Nations] Framework Convention on Climate Change [UNFCCC]. So during the annual major climate negotiations, no country can dispute any facts because they had the chance to do so when they adopted the Summary for Policymakers. In that sense, I can see the point of this process.

I think the main disadvantage of the IPCC is the immense amount of work that goes into it. Despite the fact that hundreds of scientists spend so much time on these reports, I think they get relatively little attention. I would really be interested to know how many members of the Bundestag actually read this Summary for Policymakers carefully. Most people get their information from the press in the end. So one might question the cost–benefit ratio of the IPCC reports and consider whether it would make more sense to limit them to shorter special reports. After all, the basic facts are now all on the table. Almost everything that needs to be known in order to pursue an effective climate policy is now available as verified knowledge, so, in my opinion, the IPCC has largely fulfilled its mission.

What share of your time do you spend on communication, administration, and research?
A big part of my working time is spent informing myself by reading news and technical literature. Following social media sites like Twitter is also a good source of information and keeps me updated on what my peers do. But the time I really do science, that is, write a paper or design research, is in the 10–20% range. Unfortunately, it's turned into a side job of sorts.

There are few scientists who are very much in the public eye and do a lot of public relations. Do you see science communication as a duty for scientists?
I think we have a debt to pay. Ultimately, we are funded by the public, so the public has a right to know what we are doing; we have to explain it outside the technical literature. And if there's danger ahead, then you have to warn the public as well.

That's how Prof. Drosten (a German virologist who came into national prominence as an expert during the COVID-19 pandemic) explained the state of knowledge in his podcast during the COVID-19 pandemic. I don't think there is an obligation to do this for every scientist; after all, not all climate researchers and virologists can be on the radio. That said, I wish that more colleagues would do it so that it would not rest on so few shoulders.

Of course, this is also due to our media system. The media picks out the most prominent representatives, which is a self-reinforcing feedback. As a result, the same people keep appearing in the media.

The PIK press office tries to distribute interview requests more broadly and to pass them on to young scientists as well. This prevents the same people from being interviewed all the time and gives young scientists the chance to learn how to communicate science publicly, too.

In a recent paper, you investigated what the energy company ExxonMobil knew about climate change very early on. What did you find out?
That work was about the earlier climate models that the Exxon company itself developed and ran. Back in the 1970s and early 1980s, they correctly predicted global warming caused by fossil fuels and their CO_2 emissions.

This was a paper initiated by my colleagues Geoffrey Supran and Naomi Oreskes at Harvard. Both of them are historians of science, and they asked me to help them evaluate the results of the Exxon climate models quantitatively, so I was consulted to compare the results of the climate models with observational data.

Exxon knew about the impact of CO_2 emissions on our climate early on, yet they have publicly questioned climate research to influence public debate. How strong is the influence of lobby groups from the automotive or fossil fuel industries?
Over the decades, I have come to believe that this influence is very strong. A large part of the lack of climate policy cannot be explained in any other way. Now, more and more is becoming known about how lobbying has been used to delay or prevent climate policy decisions at crucial points. The German government has, for example, also put obstacles in the way of the European Union's ambitious climate protection plans or weakened them in favor of the German automotive industry.

I consider the influence of lobbying groups to be a major reason. However, it is not always direct lobbying influence in the sense of industry leaders having Merkel's cell phone number (though that was true for some corporate CEOs), but it is also indirect lobbying influence via media reporting.

For example, this unspeakable climate skeptics debate that has been taking place in the media for decades has given the public the impression that human-caused climate change is still controversial and that there are two camps in the research community. Fortunately, this is no longer the case in Germany, but for years, this narrative dominated the debate. This was not least due to the fact that companies like Exxon, as has now been documented, funded these skeptics and their think tanks with hundreds of millions of dollars to produce pseudo-reports and arguments for climate skeptics. I consider these interests to be the main reason why we still have rising, not falling, emissions worldwide.

Do you see any reasons other than lobbying for why climate policy is so slow?
One main reason is that CO_2 emissions occur everywhere and are closely linked to our energy system. This makes everyone jointly responsible, and responsibility is diffused. That's why there are not only lobby groups that push climate-skeptic memes but also a broad public that eagerly picks them up to relieve their conscience.

Who then do you see as responsible for acting? Individuals or the state and companies?
It is certainly an interplay of individual players—some consumers who take the lead, and also politicians, who need to introduce rules that everyone abides by. Unfortunately, certain groups have managed to turn the debate into one about personal freedom, although of course no one would doubt that rules are needed to live together. Few people will make an effort to stop flying to Mallorca if their neighbor still does. That's why rules are needed, and that's why I see the government as in charge here.

Nevertheless, since the government is acting too slowly, we also need pressure from citizens who exemplify alternatives and also take to the streets to increase pressure for more climate protection.

You were able to gain some insight into politics through your work as a member of the German Advisory Council on Global Change (WBGU). How did that work out?
The WBGU meets monthly for a two-day meeting. Their work is similar to that of the IPCC, with the only difference being that, in addition to presenting the state of affairs and possible actions, concrete policy recommendations are also expected. Ministries can comment on the draft and ask questions. Finally, a report is published. And if you're lucky, a minister will come to the handover for an hour. These events have been bizarre sometimes. One time, for example, Sigmar Gabriel (former German Minister for Economic Affairs) got very cross because we had recommended ending the admixture of biofuels to gasoline, which was against the government's policy at the time. As a scientific consultant, if you say something that politicians don't like to hear, they will tell you…

Is our system at fault for the slow progress on climate policy? What would you like to see changed? Do we need peer review instead of federal elections?
No. Ultimately, elected members of parliament have to make the decisions because politics is also about making value decisions, weighing different interests, and mediation. However, it would be beneficial if scientific advisory bodies were actually taken seriously and not just used as an alibi to push through what one wants to do anyway and to ignore and reject unwanted advice.

Nevertheless, I believe that the expert reports have had a certain effect, which, of course, is difficult to measure. At the WBGU, we have realized that some expert reports have been received and reflected on much more outside Germany than within. So even if politicians don't take these reports seriously at first, they do contribute to the debate. If more and more advisory bodies across Europe recommend the end of the use of biofuels in cars, then after 20 years, it will eventually be accepted.

How important are protest movements? From Fridays for Future to the Last Generation, which groups do you support personally?
I'm more with Fridays for Future. Social researchers can certainly better answer which protest forms are most effective. In any case, I was surprised by the impact Greta Thunberg had holding up her sign in front of the Swedish Parliament. To me, that shows that the time was ripe: young people were already alarmed, and Greta's movement took off because of that. I would describe that as a kind of social tipping point. I found the global movement that Greta sparked very impressive, and it has advanced the climate debate much more than all of our IPCC reports. Nevertheless, the scientific facts are, of course, very important. Greta has read the reports herself and really understands the technical details within the scientific literature. When she visited us here at the PIK, we had a long discussion. All in all, I think this grassroots movement is really important!

What is your recommendation to young people: should they study climate science or rather put their time and energy into activism or politics?
I think that the two are not mutually exclusive. From my life experience, I would say that it is helpful to have a degree in climate science if you want to achieve something politically. Moreover, you have to assume that you're going to have to stay in the game for a long time. Even if emissions start to fall soon, the battle over climate policy decisions will continue for the next few decades, hopefully in an environment of falling global emissions. Even so, the issue will be that emissions will need to fall even faster. I think one will need a lot of stamina, and for that reason alone, one should finish one's studies instead of just becoming a full-time activist for the next three years. I think it makes a lot of sense to combine the two.

What is your response to people claiming that scientists ought to remain politically neutral?
I think that, as scientists, we have a duty to society. This critique mostly comes from people who don't want to hear our news and try to discredit it as activism.

Can we still limit warming to 1.5 °C?
Physically, it's still possible, but not with current policies. So when people say that we can't limit warming to below 1.5 °C anymore, that's a political assessment, not a physical one. I can understand that, and I don't think we're very likely to stay below 1.5 °C of warming either. But if people want to argue that the target is also physically unachievable, I have to disagree. Model simulations say otherwise.

Some people think the thermal inertia of the oceans will make it impossible. This thermal inertia exists and implies that the ocean will continue to warm even with zero emissions. But at the same time, there is an imbalance in CO_2 concentration between the atmosphere and the ocean, so the ocean will still absorb CO_2 from the atmosphere, which balances the thermal inertia effect. So if we reach zero emissions fast enough, the 1.5 °C limit is still feasible.

What would be the most pressing climate policy measures at this point in time?
I think we need a socially just CO_2 pricing scheme. The transition to renewable energies is equally important because we need to switch more and more processes to electricity: electromobility, heat pumps, and other things will require much more electricity. Renewable energies are therefore a very central issue. The transition must happen in all sectors, and we must move away from inefficient combustion processes. I don't think only a few selected measures will get us anywhere. Instead, there are many measures across all sectors that must be implemented quickly and simultaneously. There is no silver bullet, it's more like a bronze buckshot.

Reference

Caesar L, Rahmstorf S, Robinson A et al (2018) Observed fingerprint of a weakening Atlantic Ocean overturning circulation. Nature 556:191–196. https://doi.org/10.1038/s41586-018-0006-5

Chapter 4
Nico Wunderling: "The Climate System is a Series of Dominoes"

Leon Galbas and Maja Maschke

©Willem Huiskamp

Nico Wunderling (*1992) is a physicist, focussing his research on the resilience of the Earth system, the nonlinear dynamics of interacting tipping elements in the climate system and their effect on society and vice versa. He studied at the Friedrich-Alexander University of Erlangen-Nuremberg and completed his Ph.D. on "Nonlinear Dynamics and Interactions of Tipping Elements in the Earth System" in 2021, supervised by Ricarda Winkelmann. For his thesis, he was awarded the Friedrich Hirzebruch Doctoral Award 2023 by the German Academic Scholarship Foundation. He has since been engaged as a postdoctoral researcher at the Potsdam Institute

Date of the Interview: February 28, 2023

G. Lohmann, *Conversations on Climate: The People Behind the Science*, SpringerBriefs in Climate Studies, https://doi.org/10.1007/978-3-032-19147-2_4

for Climate Impact Research, specifically in the FutureLab on Earth Resilience in the Anthropocene and the Stockholm Resilience Centre.

How did you end up choosing to pursue a Ph.D. in climate modeling and how has your career gone so far?
I am a physicist by training. I studied physics at the University of Erlangen-Nuremberg. At the end of my studies, I was a bit lost. Having studied something quantitative, I wanted to use my skills to actually make a difference. While I wasn't particularly looking into climate research only, I wanted to do something meaningful that was more than just crunching numbers. That's how I ended up at the Potsdam Institute for Climate Impact Research (PIK). From the beginning, I felt very at home here and so I started my Ph.D. at the end of 2017. That's how I made the transition from physics to climate research, or climate physics.

Would you say that, as a physicist, you had an advantage or a disadvantage when making the move to the field of climate research?
With quantitative sciences, I don't think you have a big disadvantage here since you can get into the technical details relatively quickly. Of course, you won't initially know your way around as well as those who have been in the field for ten years, but you can definitely pursue a career in climate science without a particular Earth system background.

How would you describe your research interests?
I come from the fields of complex systems science, network science, and dynamical systems. With these methods, I study the tipping elements of the climate system, such as major ocean currents, the Greenland ice sheet, and the Amazon rainforest. Specifically, I look at how these tipping elements interact with each other and whether interactions between tipping elements can destabilize the climate system overall.

What does your day-to-day work look like?
In the beginning of my Ph.D., it was a lot of literature research, programming, and, of course, debugging since I rarely produce error-free code on the first go. From there, I began to approach my research questions. Now, five years later, my daily work as a postdoctoral researcher involves writing papers, giving lectures, and taking part in public outreach activities. Sometimes this is challenging, and it often includes new experiences for me as a physicist. I particularly enjoy supervising students and helping them pursue their own careers. I find that kind of work very fun and engaging. As you can see, daily work in research can be quite diverse.

You already mentioned the term *tipping elements*. What exactly are tipping elements and tipping points in the Earth system?
Tipping points are critical points beyond which certain Earth system components change qualitatively, driven by internal, self-reinforcing feedbacks. There are a whole series of them in the climate system, such as the Greenland and Antarctic ice sheets, the Atlantic Meridional Overturning Circulation (AMOC), Amazon rainforests, and monsoon systems. All these tipping elements feature some kind of critical threshold in terms of global warming, which, when exceeded, can lead to drastic changes in

these systems. In the case of the Greenland ice sheet, for example, this tipping point lies between 1° and 3° of global warming above pre-industrial levels. If this threshold is crossed, the Greenland ice sheet will disappear irreversibly.

Could you explain what processes cause tipping to happen?
Consider the Greenland ice sheet. If you walk up a mountain, for example, you'll see about 1° of cooling for every 100 m you climb. The Greenland ice sheet is two kilometers thick on average. As it begins to melt significantly and thus becomes thinner and decreases in height, the temperature at the surface becomes warmer and warmer. When enough ice has melted such that surface temperatures are predominantly positive, a tipping point has been passed and the ice sheet does not recover. This self-reinforcing feedback mechanism is called *melt-elevation feedback*.

Your research is concerned with the modeling of such tipping elements. What is it that makes your work particularly exciting or challenging?
What I really love about my models is that, despite their simplicity, it is possible to quantify existential risks arising from tipping elements. These are, for instance, a melting Greenland ice sheet or an interrupted AMOC. As such, they fill a gap as compared to more complex climate models, which either may not have included all relevant processes or are very computationally expensive so that all relevant uncertainties cannot be taken into account properly. Therefore, I feel that my research can contribute to present climate change challenges.

Among the tipping elements you've already mentioned, are there any for which the identified uncertainties are particularly large or small?
Uncertainties in tipping points are still significant. Two systems for which uncertainties are rather modest are the Greenland and West Antarctic ice sheets. According to the newest scientific literature, their tipping point lies between 1° and 3° above pre-industrial levels. In comparison, the Amazon rainforest tips somewhere between 2° and 6°. This means that the uncertainty in critical temperature thresholds is two times higher for the Amazon rainforest as compared to the large ice sheets on Greenland and West Antarctica.

Do we know why some systems' behavior is harder to predict?
There are two important factors that need to be considered. First, there is data availability. It's simply hard to directly gather the climate data that we need to initialize our models, be it long-enough data series from satellites or data from paleoclimate archives. Second, global mean temperature is not always the critical parameter that is decisive for each of the climate tipping elements. In the case of the ice sheets, the critical parameter is the local surface or ocean temperature, which can be translated back into levels of global mean temperature increase. For the Amazon rainforest, the critical parameter is moisture supply, which is largely influenced by local rainfall. However, it is less clear how this rainfall changes with increasing global temperature levels.

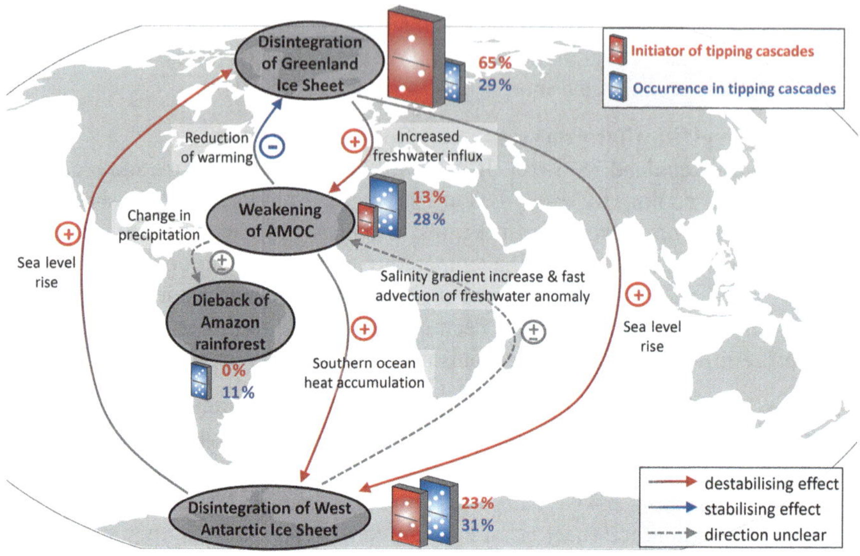

Fig. 4.1 Interactions between climate tipping elements and their roles in tipping cascades. Figure from Wunderling et al. (2021), ©Author(s) 2021, licensed under CC BY 4.0

So you're saying that change in global mean temperature is sometimes not actually the relevant quantity for tipping elements?
Precisely. A second point that is important here is that it's not just the absolute change in temperature but also how quickly it occurs. Some tipping points can tip over because the speed of warming is too fast to keep them safe.

In your publications, one often comes across the term *cascading tipping elements*. Could you explain this idea in more detail?
You can think of it as tipping elements being represented by a chain of dominoes. If you tip over the first domino, it may topple subsequent ones and lead to a tipping cascade. Our research has shown that there are different types of these dominoes, namely those that are at the beginning of a tipping cascade and those that follow. An example of a domino at the beginning is the Greenland ice sheet. When the ice sheet starts to melt, a large amount of salt-free meltwater enters the ocean. The AMOC, in particular, which includes the Gulf Stream, is sensitive to this. The AMOC is thus a subsequent domino, similar to the Amazon rainforest, which is located at the very end of such a tipping cascade. To conclude: meltwater from Greenland changes the circulation in the Atlantic, which in turn may reduce the precipitation over the Amazon rainforest, possibly sufficiently enough that parts of it would transition to a savanna state (Fig. 4.1).

You apply a network modeling approach to coupled tipping elements in your research. What does the term *network* mean exactly in this context?
A network consists of two basic components: nodes and interactions, or links. Mathematicians or physicists might be more used to the term *graph*. The tipping elements correspond to the nodes, which are subject to dynamic behavior. These dynamics can be described by a differential equation. The interactions between the tipping elements are, correspondingly, the links or edges. In the simplest case, one may use linear couplings of differential equations. Depending on how strong these couplings or links are, the closer the dominoes are to each other. That is to say, if I move one domino a little bit towards the next, the probability that I might trigger a cascade is higher. For the tipping elements in the climate system, of which there are only ten to fifteen, this network is relatively small, but there are also tipping elements, such as the Amazon rainforest, which in themselves consist of a network with interactions. Simply explained, each region in the rainforest is a mini-tipping-element, which is then in turn connected to other mini-tipping-elements in the Amazon via what is termed *moisture recycling feedback*. This is the process by which moisture falls as rain over the Amazon rainforest, evaporates again, and is thus carried further inland by the wind. That's how these mini-tipping-elements are connected. One might consider several hundred of them, depending on how finely one wants to resolve their structure.

Do you use linear couplings because that's a simple approach that works, or can you justify that physically?
First of all, it's the simplest approach. Since there are really few studies on the exact form of interactions, a good first step is to assume they are linear. Then, it is up to us scientists to take a closer look and see whether this is true or whether interactions are better represented by quadratic or other nonlinear couplings. Studies are just now emerging that look into the interactions more closely. Here, a very first step is to properly constrain the interaction strength, for instance, by using the most up-to-date satellite observations or Earth system models. In current research , interactions between tipping elements are a very large source of uncertainty; sometimes it is even unknown whether an interaction is stabilizing or destabilizing. That's why we need risk-assessment approaches powered by several million simulations that can properly take such uncertainties into account.

Are there parts of the Earth system for which network models do not work since the Earth system is not a discrete system?
It is always difficult to decide whether two systems form two different tipping elements or whether they are part of the same tipping element and thus have to be modeled together. It's not that simple at all. If you think of oceans and ice sheets, it is rather intuitive that they can be described as two separate systems. In the case of the Amazon rainforest, however, things are much more complex because you can look at it on different spatial scales. On the one hand, you could consider the Amazon rainforest as one entire area, a single domino, so to speak. On the other hand, it has been shown that this is not quite correct because smaller spatial scales are very important. It seems that the southern part of the Amazon rainforest is much

more at risk of tipping towards a savanna state than the northern part since regional precipitation patterns and other effects play a role there. In this respect, it is actually the case that some tipping elements consist of several small tipping elements that can interact with each other. The question of whether the Earth system as a whole can be seen as one big domino that either tips over, meaning all tipping elements are in the tipped-over state, or not, is subject to current research. This is the so-called hothouse hypothesis, which is definitely a controversial topic within the scientific community, and a final answer has not been found yet.

Are there interesting results on networks from completely different fields of research that can be applied to your work?
Yes, in ecology, for example, there is a whole body of knowledge about interacting ecosystems. Something similar exists in the social sciences as well: when different groups of agents interact with each other, you can study, for example, how opinion shifts or even how uprisings can emerge. Similar approaches can be applied to the climate system.

What's possible in any case is to make use of all the results from graph theory in discrete mathematics and network science, such as shortest path lengths between two nodes, the choice of weights in the interactions, and so on.

You said that you also draw on results from the social sciences and publish on so-called *social tipping elements*, too. What exactly can be understood by social tipping elements, and since when have researchers been interested in them?
If you think of it like a narrative, the climate tipping points are generally rather negative events that we as a society should prevent. At this point, we are already quite deep into global warming and this raises the question: "Which social processes can help stop climate change quickly enough or perhaps even turn it back?" Social tipping elements can be such a solution. We have seen that social upheaval processes such as the climate strike movement can arise very quickly and then lead to major social changes within a few months or years. Compared to climate tipping elements, this is extremely fast. These social tipping processes are necessary for a sustainability transformation and perhaps even a prerequisite.

Earth system tipping elements can usually be modeled well by physical laws. How does that work with social tipping elements? How can you model them quantitatively?
A typical tool for modeling society are network models. In these network models, the nodes may represent people or groups of people, and the links are interactions between these groups of people. One very famous example of social dynamics using networks is opinion dynamics. Imagine a network of different people who are friends. Let's say you have a lot of peers who all have the same opinion on something, which you yourself do not share at first. What can usually be observed is that there is something like a contagion process going on. This means that while you were of a different opinion in the beginning, the exchange and conversations with your friends led you to change your opinion in the end; you were "contagioned" with the opinion of your friends. This not only applies to opinion dynamics but to behavior in general.

This means that if the new norm is to behave more sustainably, it is much more likely that you will adopt a similar sustainable behavior. In brief, the decisive factors are the environment you live in and the people you interact with.

Could you give a simple example to illustrate how the climate and the behavior of a society influence each other?
A very simple example is the Paris Climate Agreement itself. If there were no global warming, then one would not have come to the conclusion that we must now limit global warming to a certain level. There has been a change of thinking here since the '60s, '70s and '80s up to the 2010s. Now we can at least agree that it is a reasonable goal not to exceed 1.5°–2°. From that point of view, this constitutes a feedback from the Earth system back to the political or societal level. As a side note, the Paris Climate Agreement was partially informed by trying not to exceed the thresholds of climate tipping elements, which is not guaranteed at 1.5 °C but the probability is at least limited.

Which agents play the biggest role in such a network of social tipping elements?
From my point of view, different roles are important here. Of course, we need to look at policymakers. Other than that, there is also the corporate and business sector and civil society itself. Naturally, the role of civil society varies across governmental systems, and I can't say exactly whether societal efforts play a major role in autocratic systems. Although autocracies should be able to react very quickly, this has not happened so far in the context of sustainability.

Does thinking about climate change all the time in your day-to-day work make it easier or harder for you personally to deal with the burden that comes with it?
Actually, I can say that I'm not down about it all the time because I believe that we as a society will be able to counteract climate change in the coming years and decades, hopefully to a sufficient extent. We simply have to act now. We cannot wait another 20–30 years. There is the Paris Agreement, and there are already plans for decarbonization by 2050 and before for some countries, while other countries aim for decades in the second half of the 21st century. So I have hope that we'll move into a sustainable future. I'd say my role as a scientist is to inform society, to say, "These are the things we know" and, for those who are interested, "This we don't know yet." Based on that, though, other groups in society, such as politicians and businesses, have to act on the knowledge and implement policies.

Are there any big unanswered questions in your field of research where you think, were they answered, we would see great advancements, or is it more akin to a lot of little steps that are made over time?
A very big point—we touched on it at the beginning—is the tipping uncertainties of the different tipping elements. They are almost absurdly large in some cases, between 1.5° and 6°, 7°, or even 8°, as is the case for the AMOC. I think it would help if we were to add a whole series of studies on top of what has already been done, thus constraining the tipping points. We need to know the safety zones of the Earth system. Another point of scientific action that I think is important is coupling the

Earth system itself to social dynamics. That means no longer viewing humans and the Earth system as isolated entities, which has mostly been the approach in the past. I think those are two points that would be extremely helpful in climate research.

What about the climate strike movement or other social processes? Do they have an impact on the research community in any way? Is there more funding now, or would you say that increased awareness has had little impact on research for the time being?
On this, I have to offer a somewhat negative outlook: when I came to PIK, I thought that climate research was an exciting topic that was of great social relevance. From my point of view, this is not reflected in our funding. Climate research is, in terms of funding, by no means as established as certain other fields. That's a big problem, to put it bluntly. On the other hand, it's nice to see that scientific results are being taken up and are leading to societal change. That is a positive outlook. Many PIK colleagues, including Stefan Rahmstorf, have regularly provided expert input for politicians and the general public but also the climate strike movement in Germany.

Are there other external factors that act as obstacles in the way of the research community, for example, concerning working conditions?
Working conditions are actually a big problem across the German scientific system. I would say that, until the end of the doctorate period, things are okay, but after that, it's a rocky road. Of course, some may say this is due to the absolute income compared to the private sector. More importantly, it is because of the great uncertainties in employment caused by the relatively short-term prospects that are created for researchers. It is not surprising that many German researchers go to the UK, the Netherlands, or, quite commonly, the US to spend a few years doing research there. These conditions exclude a large range of people. When you are between 30 and 40, other elements start to become important, say when it comes to starting a family. You're just not as flexible then, and I think that's a big problem for the science system itself. This definitely needs to change, not only in climate research but in the general German scientific system.

Was it difficult for you as a prospective scientist to gain a foothold in the community after your studies and especially after your doctorate? Did you have mentors who helped you?
On a scientific level, the exchange with and support of my Ph.D. supervisors Ricarda Winkelmann and Jonathan Donges has certainly been key. Other than that, a huge advantage of a research institute like PIK is that there are a lot of really high-profile researchers here, which gives you the privilege of benefiting from their knowledge by just going next door. That's great, of course, and cannot be found everywhere. In this respect, a specialized institute is a great place to dive into research more quickly. Maybe that's why, to get back to the beginning, it's easier to overcome the lack of knowledge in climate science as a physicist compared to pure Earth system scientists.

What role does the Potsdam Institute play in terms of communicating science to the general public as well as to policymakers?
Two things: first, there is a relatively small but very good media and science communication team here at PIK. Second, the institute has become an authoritative voice in climate science since its formation more than 30 years ago, so there are definitely people from politics who approach PIK and ask for expertise in certain areas of sustainability.

Would you say that, in climate research, it is particularly important to be skilled in science communication as a researcher, and should that also be part of a scientific education?
Yes, how and what you communicate is definitely important. There are training programs organized by the University of Potsdam which you can participate in, but there is definitely room to better institutionalize that. If you work in such a relevant field, public communication should really be formally included in your training as a Ph.D. or postdoc.

As a scientist, do you also come into contact with climate change deniers? Have you been, say, verbally attacked or impacted in any other way?
Fortunately, I don't notice it in my own everyday life, but once I received a call via my telephone at PIK from someone who didn't believe in climate science and wanted to convince me of that. We had a very short conversation. However, for other colleagues, especially some of our more senior researchers here at PIK, this is different; they have many more interactions with climate skeptics or deniers, usually via online media such as Twitter or Instagram.

What result of your personal research makes you particularly proud?
What really pleased me was that one of our publication results was presented at Conference of the Parties (COP) 26 in Glasgow as one of ten climate highlights that year. That is, of course, a great platform. The topic of that publication were precisely the tipping cascades that we were talking about earlier: which dominoes are at the beginning, which are in the middle, and which are perhaps at the end? The overall result was that, in terms of tipping elements, you can think of the climate system as a series of dominoes placed one after the other, which destabilize the system further than global warming alone would.

What developments in climate research give you hope for the future?
I've been noticing that we have achieved a certain social relevance, which I find very encouraging. Our former director, John Schellnhuber, once said that scientific results need 20 or 25 years to make it from research to society and to then lead to action. I think that's where we are now.

Reference

Wunderling N et al (2021) Interacting tipping elements increase risk of climate Domino effects under global warming. Earth Syst Dyn 12(2):601–619. https://doi.org/10.5194/esd-12-601-2021

Chapter 5
Tim Palmer: "After All, I Decided I'd Rather Do Something Useful"

Moritz Thies and Pablo Toussaint

©Tim Palmer

Tim Palmer (*1952) is a Royal Society Research Professor in Climate Physics at the University of Oxford. After his Ph.D. in the field of general relativity, he switched to atmospheric physics, where he pioneered the development of ensemble forecasts to estimate the uncertainty of weather and climate predictions. He has received numerous awards for his work, including the Institute of Physics' Dirac Gold Medal.

Date of the Interview: January 30, 2023

G. Lohmann, *Conversations on Climate: The People Behind the Science*,
SpringerBriefs in Climate Studies, https://doi.org/10.1007/978-3-032-19147-2_5

Professor Palmer, what question has been on your mind the most today?
Well, if you really want to know, it's something very mundane. Last week, the tires of my bicycle were rubbing. There was something horribly wrong with it, and I tried to fix it over the weekend, but I wasn't sure whether my fix would work. So the thing I was thinking about most this morning was, when I get to cycle in on the last two miles, will the things that I did to the bicycle over the weekend work? It turned out that it did fix it. Having sorted the bicycle, I then worried about a group talk I gave this morning about attributing the 2022 floods in Pakistan. So you get the sense of priority that I have in my life.

What part of your work do you enjoy the most?
Well, I know what parts of work I enjoy the least: meetings and anything to do with bureaucracy. During my time at the European Center for Medium-Range Weather Forecasting (ECMWF), I had to write a plan every year about what I would do in a 4 year period. I barely know what I will do in a month, never mind 1 year, and absolutely never mind 4 years. So writing reports where you have to say what you will do in 4 years, I just found utterly impossible. What I really wanted to do was write a report saying, "Trust me, it'll be fine." But, anyway, I like research, I like thinking, and I enjoy writing papers because you write a paper when you finish a piece of research. The trouble with reports is that you write them when you haven't finished everything. That's not something I'm enthusiastic about.

You wrote your doctoral thesis in the field of general relativity. Afterwards, you decided to change research areas and move to the UK's national weather service, the Met Office. What was your main motivation to change your research field after your Ph.D.?
It wasn't an easy decision to make because I had a job offer to work in Cambridge with Stephen Hawking. But there were two issues. The work was very abstract and didn't impact people's lives at all. And the second thing was that the area that I would say is the most important in theoretical physics is joining together quantum mechanics and general relativity, and still, there isn't a successful theory. I just got kind of nervous that if you spent a whole career working on this without really achieving anything, then not only have you not done anything that benefits society, you haven't done anything that benefits anyone. Because most of what you would've done would be pretty much a dead end. So after all, I decided that I'd rather do something useful. It wasn't that I had thought about weather and climate. I just met a guy called Raymond Hyde, whose interests span both astronomy and climatology. He convinced me that it would be an interesting field where I could apply my mathematical capabilities, too.

One of your main areas of research is ensemble predictions. Could you explain to us what ensemble predictions are and how they can help improve climate predictions?
Ensemble prediction is closely related to Monte Carlo forecasting. It means running your model multiple times to get a probability distribution. I started getting involved in this when I joined the Met Office in the late 1970s in an area that was called the

High Atmosphere Branch. A group was already doing monthly forecasting using statistical empirical models, which we'd call data-driven. They calculated probabilities for different types of weather patterns. My job was to introduce what was then called dynamic models to the long-range forecasting problem. If you want to blend dynamical information with statistical information, the dynamical information would also have to be in the form of combined probabilities. So, it was clear that we would have to run some kind of Monte Carlo system to produce probabilities. We managed to set up such a forecasting system in 1985, which was the world's first operational real-time ensemble forecasting system for monthly predictions. But there was a lot of resistance to this idea because of a paper by Joel Charney. They studied how far one could make deterministic forecasts and came up with this timescale of 2 weeks. People interpreted this to mean that predictions on a shorter time scale should be done deterministically, and after 2 weeks, they should be done probabilistically. An ensemble for monthly forecasting is deterministic and above the timescale of 2 weeks. But this limit is a misleading concept. It's a statistical average, if you like. There will be many situations where the actual predictability of the day is much less than 2 weeks, and there'll be occasions when it's greater than 2 weeks.

In 1987, there was this very famous storm in the UK. The BBC weather forecaster came on television that day and said, "It's going to be a nice day, no problem." And the worst storm in 300 years happened. It was catastrophic. The media said that these meteorologists were completely useless and that the Met Office director should resign. Retrospectively, we showed that this was an excellent example of a situation where the predictability was much less than 2 weeks. This made people realize that if we want to avoid this type of media criticism, we've got to do something about forecast predictability. And the ensemble was the only serious game in town to do it. Any week now, the ECMWF will release a new cycle of the ensemble, which will run at the same resolution as the deterministic forecast. The ensemble is now the primary forecast tool. For me, it's a major landmark in my career because I hardly believed I would see the day when that would happen.

Could you give an example of how those ensemble probabilities can be used to make better policy decisions?

The one thing that I find extremely encouraging is how disaster relief agencies react to forecasts of extreme events. In the past, the problem was that the agencies only reacted after a severe event had already happened. Imagine a hurricane hitting an island. If the agencies become active afterwards, it's really difficult to bring food, medicine, and water to people because the hurricane has devastated the infrastructure. You could ask, "Why didn't they act before, based on the forecast?" If you only have a deterministic forecast, particularly of an extreme event, it is very unreliable. The disaster relief agencies have a limited amount of money, so if they go in whenever there's some hint of an event, they quickly end up bankrupt. Thus, they need a more objective way to decide when it makes sense to become active. An ensemble prediction gives them that objective criterion. They can do a cost-benefit analysis ahead of time. Based on their budget, they could say it makes sense to be proactive if the probability of a disaster exceeds some predetermined threshold. That is radically changing the way in which humanitarian disaster relief agencies can work.

You already mentioned this threshold of 2 weeks in weather prediction, but climate prediction can be made decades into the future. How can you explain this difference in weather and climate predictions to a layperson?
Just think about the annual cycle. If I'm in Oxford in January, I don't know exactly what the weather will be like 6 months from now, but I can say with a pretty high degree of certainty that it'll be warmer than it is today for the simple reason that every day the Sun is getting a little bit higher in the sky and the days are getting longer and longer. So, there's a very predictable external forcing, creating a predictable signal on top of the chaotic weather fluctuation. Climate change is not different from that. Here, the predictable signal is our emissions of carbon dioxide. Therefore, the climate question is not an initial value problem. We do not take today's weather and calculate what the weather 6 or 7 days from now will be like. Instead, we look at how the weather statistics change if we increase the carbon dioxide concentration due to our fairly predictable carbon dioxide emissions. In that sense, the climate change problem is more like predicting the effects of the sun's annual cycle, but over a much longer timescale.

You said that we might have uncertainties due to the used models. But while measuring, we can also get uncertainties from noise. Traditionally, noise is considered disruptive, and its effects are kept as low as possible. To what extent can noise also be used positively?
Well, the whole thing started when we originally introduced the ensemble prediction in 1992, a time when the perturbations were only in the initial conditions of the model. But we were missing the uncertainty coming from the model equations. A model is a simplification of some underlying mathematical equations, like the Navier-Stokes equations. With this simplification, you get a random error. It may have a systematic component, but the source of the error is sort of random. I gave a talk in 1995 saying that we should make our parameterization more stochastic to account for the fact that there is this random error in the truncation of the equations, and then our group sat down and came up with a practical scheme for introducing stochastic perturbations. That improved the scores, particularly in the extratropics. All that convinced me that stochasticity is a good thing, not just for straight ensemble prediction but for running the model and getting systematic errors as small as possible. Moreover, I started looking into other areas where noise can be beneficial. This goes back to Alan Turing and his question of how to construct a machine that can imitate intelligence. He said you should be introducing some kind of randomness into the model.

Another idea is whether our computers should have noisy chips in them. It takes a lot of energy to make computer bits reproducible. But couldn't that energy be spent better on other calculations rather than making the computer bits more reproducible? Furthermore, I was thinking about what physical systems actually make use of noise, and that got me thinking about neuroscience and our brain. There's no doubt that the brain is a noisy place. The brain does what it does on 20 W. It processes as much information per second as a supercomputer does when you feed it a climate model.

But it does it at 20 W rather than 20 MW. The brain does its data processing with an extraordinarily large number of neurons—80 billion neurons—but using only 20 W, which makes the brain a noisy place. The question is whether the brain constructively makes use of that noise or whether it's just a nuisance. And I think there are pretty good arguments to suggest that it does.

This shows that thinking about climate is not your only interest in science. But what does a typical day look like as a climate modeler? When do you have time to think about other areas than just the climate?
In my early days, I spent a lot of time writing code, which I guess everyone has to do. But I have to confess, I've done virtually none of that in recent years. So my typical day is not the same as 40 years ago. I probably come to Oxford 3 days a week or so, where I try to talk to people in my group and bounce ideas off them. The other 2 days I work at home, maybe I have a meeting. I have to go to London to see the Royal Society or various other things like that. But I certainly try to have a day at home where I either write a paper or try to think about articles. I'd say that the work I did on the brain has really emphasized to me the importance of leaving yourself enough time in the week to just relax. I'm a great believer that having big ideas doesn't happen when you are just staring at the screen trying to code some new part of the model. You've got to allow yourself time to go for a walk or a bicycle ride. Whether you like it or not, your brain will sort of churn over things that you've been working on during the week—and that's when ideas happen. I try to ensure that I've got a day at least, maybe spread over several hours in the week, where I'm not doing anything. I know bosses don't like the idea of you spending time not doing anything. You've got to have timesheets for everything that you do. But the more timesheets you have in life, the less likely it is that you will get breakthrough ideas. In *Nature* there was an article a few weeks ago about why there seemed to be much fewer breakthrough ideas today than maybe 50 years ago, and I think one of the reasons is that people just fill up their calendars with meetings and don't have any time to do anything.

Climate research is always a political issue. In which areas of your work might ethical considerations come into play?
Well, that's the whole thing: any work that a climate scientist does has political implications. I give many outreach talks where people want to know, "What can I do about climate change?" I certainly don't steer away from those types of discussions, but on the other hand, it's important to be able to clearly explain the difference between science and things that involve value judgments. Is carbon dioxide a greenhouse gas? Well, yes, carbon dioxide is a greenhouse gas; that's a scientific question. Will our emissions of carbon dioxide increase the temperature of the planet? Yes, it will increase the temperature of the planet. That's a scientific fact. Will it increase the likelihood of extreme weather? Yes, again. So should we cut our emissions of carbon dioxide? That's a question that involves value judgments. I think if a scientist says, yes, we must cut our emissions of carbon dioxide, then it's essential for people to understand that they're now speaking not as a neutral scientist but as somebody who's put on a particular political or environmental hat. One person might be better at making value judgments than others. When I give talks, I can explain to you as

best I can what I think the science behind climate change is. But when people say, "Listen to the scientists," as if scientists tell you unambiguously that you've got to cut emissions, then I think it can't be the right way. The other thing is that science is uncertain. Anybody who says climate change will be a catastrophe, or conversely, it will be nothing to worry about, is scientifically incorrect.

After your Ph.D. thesis in general relativity, you switched to climate physics. But 10 years later, you got back to fundamental physics again. Since then, you have published articles in both research areas. How do you manage to stay so versatile as a scientist?
I've stayed versatile by not taking jobs that would have had a significant bureaucratic component. I enjoy running with a small group of people, typically up to 15 or 20. I've deliberately avoided bigger groups, for example, running a laboratory or a big institute where you've got hundreds or even thousands of people. I don't think I'm particularly well suited for these kinds of jobs. Other people could do those jobs just as well as me, if not better. My strengths lie much more in thinking through problems in basic science. So the answer to your question is that I've just made sure to have jobs that give me enough time to do other things. If I'd been running a lab with a thousand people, there was no way I would've had any time to think about quantum mechanics. That's why, when this Royal Society Research Professorship came up in 2009, I just said: Look, this is the perfect job for me. That will take me right through my retirement. It's been fantastic, and it's allowed me to carry on working in any weather or climate, which I'm very happy to do. I've got things that I still want to do there, but at the same time, I can get back to my roots and apply some of the ideas, for instance, the geometry of chaos.

Is there a result of your research that you are particularly proud of?
I'm proud of the whole ensemble prediction work because it's affected every weather service and climate prediction worldwide. Everyone uses ensembles. I'm not saying that I was the only person to do it, but I was there at the beginning, and it's been very gratifying to see how that's become used.

As a final question, what developments in climate research give you particular hope?
Hope? Well, I can tell you what I think is needed. What keeps me awake at night in the climate research area is that our climate models can't keep pace with what's happening in the real world. In the last 2 or 3 years, we've seen some pretty extreme weather events, like the floods in Germany and Pakistan or the great heat waves in Canada. Even the UK hit 40 °C last summer. But most of these extreme events lie outside the range of traditional climate models. As a result, I think we're going into a bit of a wild west where people can say whatever they like about climate change in terms of how catastrophic it might be. The models are not providing a good check on these statements because we know that they just don't simulate extreme events very well. Personally, I believe that we've quite urgently got to do something about this. I think this can only be solved through enhanced international collaboration, like a CERN (*Conseil européen pour la recherche nucléaire)* for climate science,

that develops models with much, much higher resolutions than we currently have. For that, we need dedicated exascale computers. That will answer many questions. Right now, governments find it really hard to quantify how their country should be adapting to climate change. Will climate change manifest itself in terms of more droughts and heat waves? We're still surprised every time an extreme event happens, so we need a big push forward in climate predictions. It's undoubtedly one of the things that I strongly feel we need to work on.

Chapter 6
Axel Kleidon: "You Can View the Earth as an Onion"

Alexander Saal, Lukas Kalvoda, and Arnulf Kung

©Rudolf Wernicke

Axel Kleidon (*1969) leads an independent research group on biosphere theory and modeling at the Max Planck Institute for Biogeochemistry in Jena and is a private lecturer at Friedrich Schiller University. After his studies in physics, he started to work on the climate system and how it is influenced by vegetation during his Ph.D. Later, he developed a holistic model of the Earth system based on thermodynamics, which allows us to better understand the role of life and humanity on our planet.

Date of the Interview: March 10, 2023

G. Lohmann, *Conversations on Climate: The People Behind the Science*,
SpringerBriefs in Climate Studies, https://doi.org/10.1007/978-3-032-19147-2_6

How can one understand the Earth's climate system from a cooking pot?
In general, I think it is important to develop a conceptual understanding of what is actually happening in a system. That doesn't mean that you consider more and more processes with increasing resolution, but rather that you understand the basic principles. I like to use the cooking pot as an example from everyday life because there are two adjustment variables with which you can regulate its temperature. One is how hot the stove plate is, and the other is whether you close the lid or not. This is a very nice analogy for the Earth's atmosphere: its temperature is determined by the warming due to solar radiation and by the greenhouse effect. If the greenhouse effect increases, then you're closing the pot lid more.

On the one hand, you investigate heat fluxes, and on the other hand, you also deal a lot with entropy fluxes. Can you briefly explain what entropy is?
It is perhaps a little confusing that there are many different definitions of entropy that are used in different contexts, such as in information theory. However, I am really concerned with the physical picture of entropy, which first appeared in the 19th century when steam engines were developed and thermodynamics came into being. At first, entropy was a purely empirical quantity that could be used to determine the efficiency of a heat engine. Later, it was extended by Boltzmann's model of a gas. This describes a gas as a collection of many particles that collide with each other. In this picture, entropy helps to relate the microscopic energy distribution of individual particles to temperature on the macroscale. This concept was then further developed by Planck, which isn't widely known. He applied it to radiation and considered light as a flux of particles, which we now call photons. Entropy always comes into play when we describe things at the macroscopic level and don't want to have anything to do with quantum physics, meaning the microscopic details. To do climate physics, for example, we don't need many quantum mechanical and microscopic details.

Which role does entropy play in the climate system?
Entropy is a really important physical quantity! If we look at what drives the Earth system, that is, what constitutes its dynamics, it is work. In the atmosphere, for example, work has to be done to keep air moving because otherwise friction would eventually slow it down to a halt. The water cycle also involves work. This work is performed by thermodynamic processes that are driven by radiation and differences in heating—for example, between the Earth's surface and the higher atmospheric layers—and can thus be described as a heat engine. Therefore, the dynamics of the Earth system are also subject to the corresponding thermodynamic limits. These ultimately result from the concept of entropy.

How does your perspective on the Earth's climate system differ from that of other climate scientists?
Actually, it is not a fundamentally different picture, but it is a completely different perspective because it focuses centrally on thermodynamics, energy turnover, and its limits. The assumption I work with is the maximum power principle: The Earth system does as much work as possible according to the laws of nature. This can be translated into equations and be used to make very simple and physically sound

statements, for example, on climate change. Complex climate models often use semi-empirical approaches, especially with regard to so-called dissipative processes, or friction. This is similar to physics in school, where friction is always somewhat swept under the carpet. Besides, friction takes place at the surface, which is less interesting for meteorologists than the middle atmosphere anyway. However, actually, friction is decisive because, like in a cooking pot, temperature is determined not only by the rate of warming but also by how well the system can cool down. So, the movement in the climate system is a balance between forcing and friction. The insights that emerge are consistent with more complex climate models. With very simple models based on that maximum power approach, we can explain, for example, climate change, the fact that land masses warm more than oceans, and the sensitivity of the hydrological cycle. I think that's wonderful because it's not based on the complexity of the model but on the understanding of fundamental laws of nature.

Can you explain the maximum power principle again in more detail?
It means asking the question, "What is the maximum amount of motion that the atmosphere can generate?" This maximum power is first of all limited by the Carnot efficiency of a heat engine. However, in the Earth system, there is a second limiting factor. In contrast to the heat engine, the boundary temperatures are not fixed. As the atmosphere transports heat, for example, between the surface and the atmosphere, it also reduces temperature differences. However, a heat engine is less efficient for smaller temperature differences. Consequently, the more heat the atmosphere transports, the lower its efficiency. From a physical viewpoint, this leads to a maximum power of atmospheric motion, which can be easily determined. There seems to be an overarching trend for the Earth system to perform work at its maximum capacity. This can be seen empirically not only in atmospheric motion but also in the hydrological cycle or in biotic activities, among others. This is well documented in studies. It can be used to describe the large-scale heat distribution between the tropics and the extratropics, as well as between the surface and the atmosphere. The estimates you get from that are really good.

Was there a key event that made you decide to take a new path in research?
Yes, there was. Already as a doctoral student, I was interested in James Lovelock's Gaia hypothesis. According to this hypothesis, the entire Earth can be regarded as a kind of "superorganism." This organism regulates its own development in such a way that life is not only made possible, but its environmental conditions are also continuously improved. Later, as a young postdoc, I attended a conference of the American Geophysical Union on this topic. At the conference dinner, I sat next to two very inspiring scientists. At some point, the conversation turned to the fact that radiation has different entropies, but it was unclear to us exactly how this works with solar radiation and radiation from the Earth. That was such a key experience for me because I had never thought or heard about it before. As a physicist, I thought it should be obvious which radiation has the lower entropy, but I couldn't answer it either. Shortly afterwards, at another conference, I saw the poster of a colleague from Japan, Hisashi Ozawa, who worked with maximum entropy production. That and Lovelock's work motivated me to explore the principle of maximum entropy

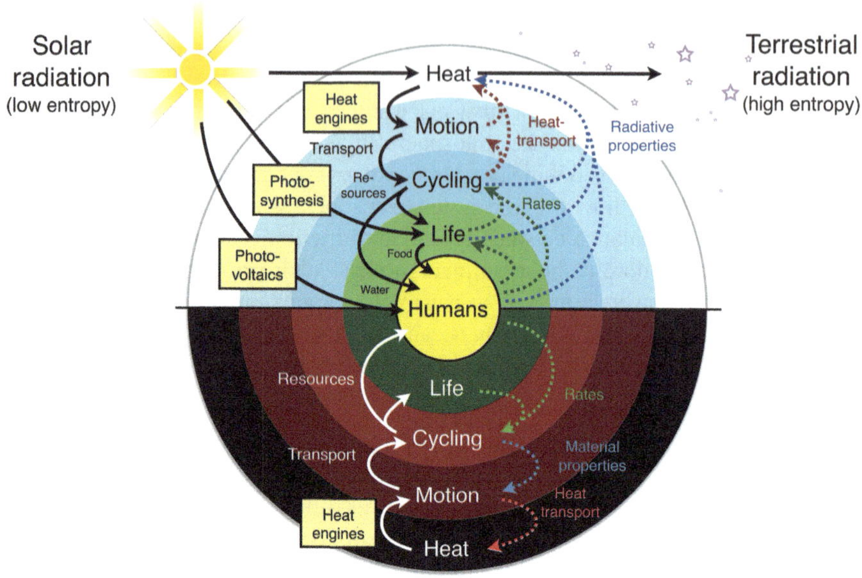

Fig. 6.1 The different shells of the hierarchy of energy flows in the Earth system. Figure from Kleidon (2023), ©Axel Kleidon 2023, licensed under CC BY 4.0

production further and use it to formulate the Gaia hypothesis more precisely. As a young scientist, it wasn't necessarily the best subject because some established people thought the principle of maximum entropy production was rubbish. When you're young, you're already intimidated by that. By now, I am more of the opinion that there is a widespread lack of understanding of it.

In one of your key articles from that time, "Life, Hierarchy, and the Thermodynamic Machinery of Planet Earth," you described the Earth as a "coupled, hierarchical and thermodynamic system." Can you elaborate on these ideas?
By hierarchical, I mean what I like to describe with an onion: you can view the Earth as an onion with different shells representing different forms of energy (Fig. 6.1). There is a transformation of energy between the shells. For example, you need radiation differences to create atmospheric motion, which is wind. When the wind sweeps over the ocean, some of the energy is transferred into the formation of waves and the mixing of the surface layer. So the energy that creates atmospheric motion carries over into ocean currents. In the same way, you can see the energy associated with the water cycle being used to transport sediments in river systems. The degradation of these then affects the continental crust over thousands or millions of years. So these processes are always coupled and also have feedback effects on each other. For example, radiation differences lead to warming differences, which in turn lead to air movement. But when air movement transports heat, it reduces precisely these warming differences.

Let's take another look at the Gaia hypothesis, according to which the Earth can be regarded as a kind of "superorganism." How does your holistic model of the Earth system relate to the Gaia hypothesis?
Interestingly, my model yields a picture that is actually very similar to the Gaia hypothesis but more quantitative and testable in principle. The heat engine, in the form of physical Earth system processes, performs work and thus drives air movement, water cycling, and so on. And life performs photosynthesis, releasing chemical energy. This energy can be used to chemically transform the Earth, which has happened continuously throughout the Earth's history. For example, the emergence of life and photosynthesis oxidized the Earth's surface. Then, oxygen accumulated in the atmosphere. Later, new life forms could make better use of this oxygen, and so on. Because life is based on CO_2, or carbon, the greenhouse effect has also radically changed during the Earth's history. This leads to changes in temperature and radiative forcing, the boundary conditions of the power plant that keeps the climate system going. On the other hand, life depends strongly on the climate system. In this respect, one can already imagine that life has changed this power plant in such a way that it itself can operate at maximum power. Thus, the Earth system is not about regulating the temperature or the wind speed but about maximizing the flow of energy.

So ultimately, basic thermodynamic principles ensure that the Earth seems to be constantly evolving like a "superorganism"?
In a sense. This metaphor of the organism came from discussions Lovelock had with Lynn Margulis, a microbiologist. Among biologists, the term "superorganism" is met with rejection, sometimes for a variety of reasons. As a brief anecdote, when I started as an assistant professor in Maryland, I was also associated with a program for ecologists. At a coffee break, I had mentioned, "Yes, I'm also interested in the Gaia hypothesis," and the response was, "Ah, from those guys who smoked too much pot in the seventies." I think this is a common attitude among biologists. In part, this could be due to the terminology of "superorganism," but a very different and important aspect is the systems approach. To put it a little more pointedly, a biologist is interested in one organism only. However, one must look at the totality of the Earth system to arrive at a statement like the Gaia hypothesis. Living things change their environment, thereby changing the boundary conditions for the physical system. This, in turn, results in a change in boundary conditions for the biosphere and thus for life. These feedback couplings act on very long time scales, but on an Earth-historical basis, of course, they eventually come through.

As you described, your models are rather conceptual in nature. What motivates you to look at these rather simple models and not follow the same path as others who make their models more and more complicated?
I do not like making models more complex. Of course, more complex models have a benefit, namely that they can make much more detailed predictions. On the other hand, with conceptual models, one can better understand which processes play a major role in a system. In fact, it often turns out that not all the complexity has to be incorporated. I would like to add that some conceptual models are also fantastic in their predictions. For example, we have a model for the potential of wind energy.

In this model, we place wind farms in the North Sea and examine how much kinetic energy is contributed by the atmosphere. How much is lost through friction? How much kinetic energy do the wind turbines extract from the wind? How much do they convert into electrical energy? From observational data on wind speeds, one can then do a simple spreadsheet calculation. This gives you estimates that are more or less as good as those from a highly complex weather prediction model. That illustrates very nicely and vividly that the most important process was identified, including the relevant feedback. The more turbines there are, the more kinetic energy is removed, and therefore the more wind speed is reduced. That sounds trivial, but it's not.

So, your core motivation is that understanding is paramount?
Yes, exactly. I think there is a huge gap in climate research because it's very one-sidedly focused on "the higher the resolution, the better" rather than saying, "Let's do just the opposite." Einstein, after all, said that a theory should be as simple as possible but not simpler. That's exactly what climate models don't give us. They go in the opposite direction.

For some years now, there has been a discussion about whether the changes to the Earth system caused by humans are so large that we find ourselves in a new Earth-historical epoch, the so-called Anthropocene. What can you say about the term "Anthropocene" from the perspective of your research?
There is a very clear conclusion. The Earth receives plenty of energy from solar radiation, and humanity consumes only a minimal amount of it. But we don't feed on heat, we need so-called "free energy" to maintain our metabolism or perform work. When we look at how much free energy we use, our energy consumption is much more comparable to Earth system processes. The power of atmospheric motion is about 1,000 TW, and the wind-driven circulation of the ocean is on the order of 5 TW. If we compare that to 18–20 TW of human energy consumption, we see how tremendous our role in the Earth system really is.

We now want to delve into what sustainability can mean in this context. In what ways could humanity's activities benefit the Earth system?
So far, we have been taking energy away from the biosphere through agriculture in order to ensure our food supply, and, in principle, we are also taking energy away from the climate system, such as through wind turbines, dams, and so on. Therefore, there's less energy available for natural processes, which means, for example, that winds become a little weaker. As a result, the available wind energy for human use is limited.

That sounds a bit depressing at first, but there is already another option nowadays: photovoltaics. Photovoltaics generates energy from sunlight but much more efficiently than photosynthesis. It doesn't need CO_2 or water and simply exports the energy as electricity. With photovoltaics, we can generate energy directly from sunlight without taking it away from the environment. As a result, we can make the Earth more powerful overall. For example, we can use the generated energy to desalinate seawater through membranes, which requires much less energy than the natural processes of evaporation and condensation. The water could then be used to expand the productivity of the biosphere in areas that are currently not productive, for example, by greening deserts. So, through some forms of technology, human activity could become more sustainable.

Are there limits to the potential of different forms of renewable energy?
Yes. Ultimately, you can think of it in hierarchical terms. You have the greatest potential if you use the Sun directly. Heat differences on the Earth's surface lead to wind, so wind has the second highest potential. Wind generates waves, among other things, but then there's not much power left. There are ideas to use ocean currents, but their potential is even smaller. It is quite clear: even in Germany, solar energy would have by far the greatest potential. At the beginning of one of my lectures, I always ask students, "How do you assess the various forms of renewable energy? How much power can they theoretically produce?" Solar energy is never rated highly, but at the end of the semester they know better.

The expansion of renewable energy can also have different effects on the Earth system. For example, if you place photovoltaic modules on a field, the field may be less productive afterwards. You have reduced the productivity of the biosphere. Instead, you could use it to make an already sealed area, like the roof of a house, more productive.

On the other hand, the more wind energy we use, the more we take away from the atmosphere. Especially regarding offshore wind energy, I now take a very critical stance because it leads to less friction at the ocean surface. Less friction means less wave formation, and less wave formation means less mixing of the ocean. But this mixing is essential to ventilate the deeper ocean layers with oxygen, supply them with nutrients, and maintain their productivity. So, in a way, we are biting the hand that feeds us.

Would you say wind power has such clear disadvantages compared to solar power that the expansion of wind turbines should be stopped immediately? Or do they still have a right to exist?
Especially in the medium term, wind power definitely has a right to exist, especially on land. The current German expansion target of 160 GW of power output from onshore wind energy by 2040 would only take out about 2% of what is lost due to friction over Germany anyway, so it's not much. The difference to offshore is that the offshore areas of Germany are much smaller, and it is planned to install wind turbines at a much higher density there. In addition, the dynamics below the sea surface are mainly wind-driven, whereas that is not the case with ground processes on land. Therefore, on land, the effects of wind energy withdrawal are not nearly as great. Furthermore, wind power generally complements solar power because it is also productive in the winter, while solar power is mainly strong in the summer. Clearly, wind energy is also much more sustainable than any form of fossil energy. At this point, I want to note that I am not focused on the technical implementation of renewable energy but on its impact on the Earth's climate system. I think this holistic view has its merits but often comes up short. From an Earth system perspective, photovoltaics is the most promising renewable energy source in the long term, whereas wind power is worthwhile as a transition technology, especially in the medium term.

We want to briefly address another topic that perhaps very few people have in mind: climate science in relation to other planets, namely astrobiology or astroclimatology. You helped develop a scale that divides exoplanets into five different classes based on their free energy footprint. To what extent can this occupation with exoplanets help us understand the evolution of the Earth's climate system?
That's a good question. Our understanding of the Earth system is based on a sample of size one. However, life may not be that special. That's more of a philosophical debate right now because no one has any evidence for it. I find it quite fascinating and have been working on it with an astrophysicist from the US, Adam Frank. We were wondering if climate change and global warming might not be anything special at all. Maybe it has already happened on other planets that are inhabited by higher life forms. All life forms that we know of require energy, and that energy has to come from somewhere. Therefore, I think our development is a typical phenomenon. We are currently overusing fossil fuels. It's like a savings account, which you exhaust extensively without creating an income for yourself. It would be exciting to see signs of this on other planets as well. What I find frustrating is that we may never know.

With your blog Earthsystem.org, through articles in *Physik in unserer Zeit* (Physics in Our Time) or via Twitter [now X], you provide information about your research results or other findings on climate research. What motivated you to get involved in science communication?
One of the editors-in-chief of *Physik in unserer Zeit* contacted me regarding my paper on the hierarchy of the Earth system and asked me if I would entertain writing an article for them. While writing it, I realized that it's nice to put things a bit more simply and thus reach a broader audience. I think it's important to communicate what you've learned from your own research since we're all paid through tax money. In May, for example, I'm going to a teacher training course in Weimar organized by Harald Lesch. The aim is to develop concepts that can be used to better include climate change and the energy transition in school education. I think it's important to start teaching these simple insights as early as possible.

Why did you decide to start your scientific career in climate research?
As a teenager in the 80s, I thought computers were totally cool, and I wanted to study something with robots. I started with electrical engineering, and it was awful. After one and a half semesters, I was at a school party and met my former physics teacher, who had at one time started a doctorate in meteorology. I told him, "I'm quitting and switching to physics." He replied, "Axel, that's what I expected." So I studied physics and minored in meteorology. In the US, I got my degree in physics and actually wanted to go in the direction of theoretical solid-state physics. I already had a Ph.D. position in Germany lined up, but then I took a course in climate modeling because meteorology used to be my minor. On the one hand, it was cool to work with computers, but perhaps even more important was the very motivating and dedicated professor who taught it. So I thought maybe I could do my Ph.D. with the meteorologists in Hamburg, and I enclosed a letter of recommendation from the professor with my application right

away. The answer was, "When do you want to start?" That shows what a significant difference inspiring individuals can make.

Have you faced, or do you face, obstacles in establishing your unconventional approaches in the scientific landscape?
Yes, on a regular basis. In that respect, it's not helpful to make things simple and do things differently. Everyone knows that you can't understand a sophisticated new climate model in detail. However, reviewers will never admit they don't have a clue, and therefore they will not criticize the approach. On the other hand, if you want to publish something simple, the reviewers understand it and are skeptical of new ideas at first. That's why it's sometimes difficult to get my work published. On the other hand, I have learned to communicate ideas better. One recognizes where the catch is in the story, where thought patterns are perhaps wrong or some assessments are incorrect. Accordingly, you can take precautions when writing. By now, my approach has been taking root more and more. At the end of the month, I'm going to India again for two weeks to give a whole series of talks on the subject.

Which result of your research are you most proud of?
I am particularly proud of my work on offshore wind energy. This is basic research with a direct practical application, involving billions of euros per year. It also shows how important basic research is. Twelve years ago, when we asked ourselves what the maximum amount of wind energy that could actually be generated was, it sounded far too abstract for others to find it interesting. At some point, we were brought together with physicist Jake Badger's group from Denmark at an "Agora Energiewende" (a German thinktank for the energy transition) workshop. This developed into a very productive collaboration. The report produced at Agora Energiewende even had an influence on the planning of Germany's offshore areas. It also reached the economic world and had a significant impact. Shortly before the publication of our report at that time, the world's largest offshore wind farm operator had reduced its output forecasts. They said it would be necessary to take into account that wind farms, after all, also reduce wind speed. This could be a coincidence, but I had met with their scientists in Hamburg a few months before.

What development in climate research gives you hope?
One positive aspect is that curiosity is also huge outside the discipline, both in academia and among the general public. Meanwhile, people are also becoming aware of the existence of man-made climate change based on their own experiences. There is a change in thinking, and I don't think that has much to do with climate research per se or with higher-resolution models. The critical point is communication, because the concepts and knowledge are there. Many complex systems can be understood with simple conceptual models. These models, I think, are best suited for communicating climate research to the public.

Reference

Kleidon A (2023) Working at the limit: a review of thermodynamics and optimality of the Earth system. Earth Syst Dyn 14(4):861–896. https://doi.org/10.5194/esd-14-861-2023

Chapter 7
Sandy Harrison: "Climate Models Assume That Plants Think Like Physicists—And Plants Don't"

Lena Hilf and Johanna Schneider

©Sandy Harrison

Sandy Harrison is Professor of Global Palaeoclimates and Biogeochemical Cycles at the University of Reading. She started her career in Cambridge, focussing on past climates. She joined a research group at the University of Wisconsin-Madison, where she worked with some of the first climate modelers, including John Kutzbach. While the climate was primarily considered in terms of the atmosphere back then, Harrison began focusing on the terrestrial biosphere and its interaction with climate over the

---

Date of the Interview: January 10, 2023

G. Lohmann, *Conversations on Climate: The People Behind the Science*, SpringerBriefs in Climate Studies, https://doi.org/10.1007/978-3-032-19147-2_7

course of her career. She is also involved in model evaluation and paleoclimate research.

Do you remember the moment you realised what the current climate change implies for the world?
I got involved in climate change from a palaeoclimatic perspective, with the idea that the climate is always changing. When we started to notice that human activities have a huge impact on the climate now, it was not a surprise. We then realised that the only way we could understand this climate change was by using climate models. If you take a long-term perspective, you see that the climate changes that we’re experiencing now are not unusual. The cause of them is unusual, but the types of climate changes are not. To put perspective on this: should we be worried about climate change? The answer for the planet is no; from the point of view of humans, yes.

What is, in your opinion, the most important topic in climate research that is currently poorly understood?
The role of plants in the climate. We started a project last year where we tried to build new models of the terrestrial biosphere and its interaction with climate. The reason I became involved there is that the terrestrial biosphere is (A) where we live and (B) where we get most of our food from. It has not been well studied by the climate modelling community. Most of the model components we use at the moment are concentrated on physics. Plants are modeled based on physics as opposed to based on ecology. This is one of the areas we really need to work on.

We have many different species on Earth—is it possible to represent this in climate models?
We have a winning ticket because evolution is operating. Yes, we have a huge biodiversity, but the rules of where which plants grow are defined by ecology and evolution. The idea is that plants already grow in places that they are adapted to. If you have small climate changes, plants will adapt to that by changing their physiology. If you have larger changes, they will adapt by changing the season over which they grow. If you have really large changes, they will adapt by migrating.

One of the approaches is to simplify the very diverse biosphere by assuming it is already pre-adapted, and we just have to ask the question: what is it adapted to? What elements of the climate system, in the environment they are living in, are the plants adjusted to? I think that simplifies things quite a lot.

In a physics-based approach, you want to have the physics of every process for every single species in place. Our approach says the plants already know what’s good for them. That’s an exciting breakthrough that has happened in the last few years—realising that Darwin got it right.

How can we reduce vegetation to certain plant traits?
One of the most important plant traits in terms of climate is the trade-off between taking in carbon and losing water, called stomatal conductance. Plants have to take in CO_2 from the atmosphere to grow. By doing that, they have their stomata open, which means they are losing water. So, depending on what their environmental conditions are, you have to balance those two needs. When you are in a very dry place, you

want to protect the water. If you are in a wet place, you don't need to protect your water, and you can leave your stomata open. That key trait of stomatal conductance is something that depends on the interaction between plants and their environment.

How much do we neglect when we reduce the biosphere to a few plant traits?
It depends on your question. If all we are interested in is how much photosynthesis is going on, then you can have a really simple model. If you are interested in changing albedo, you need a slightly more complex model with more categories. You need evergreen trees versus deciduous trees and understory plants. There is nothing like "six traits define the world," but for each process, you can probably come down to a small number of categories that you're interested in. The question is, as you put those processes together, how many categories do you end up with? This is something we haven't solved yet, but analysis of plant trait data suggests that we don't need to represent every single species unless you are interested in biodiversity itself. We analysed plant data on the shape and function of leaves, for example. Often, two plants will adopt completely different sets of traits to achieve the same goal. Think about plants in very arid environments: some of them cope with this by having very small leaves, and some of them cope by having no leaves at all. There are lots of strategies that plants can use to adapt to the same environment. This is presumably the root cause of the very many species that we have.

What do you think are the most important mechanisms in climate that are impacted by the biosphere?
The first one is land surface albedo, which will impact how much radiation is absorbed and how much is reflected into the atmosphere. This is very much controlled by the amount and type of vegetation. The second one has to be the carbon cycle: the uptake of carbon and the release of carbon into the atmosphere are influenced by the plants that are there. The third one is water exchange. Plants are really important in controlling those exchanges and how they happen between the land surface and the atmosphere.

How important is biodiversity for the climate, and how does it affect it?
Biodiversity itself probably isn't important for climate, but it's important for how plants deal with the changing climate. We have such a biodiverse universe because organisms have to adapt to short-term variability, long-term variability, and very-long-term variability. There are lots of strategies for how plants could adapt, and they probably vary depending on how big the natural climate variability is. For me, biodiversity is a *response* to the climate. In terms of energy, water, and carbon exchanges, you could probably have pine forests everywhere, and it wouldn't be such a bad thing. But obviously we want to have a biodiverse terrestrial biosphere because it has lots of resources and is more resilient to climate variability.

How do vegetation, climate, and wildfires interact?
They all impact one another. Vegetation is the fuel load, and depending on the kind and amount of vegetation, there are very different fire regimes. In savannas, you have a lot of ground fires because there is a lot of grass. In eucalypt forests, you have crown fires because there is a lot of woody material that can burn. The climate

can impact this by drying out the vegetation. If you have very dry conditions for a short period of time and then it rains, you will stop any fires that are there. Fires have a direct impact on the climate because they emit gases and black carbon into the atmosphere. By destroying the vegetation, they change the albedo, and by putting ash and other materials on the soils, they change the hydrology. If a fire goes through, it is going to regenerate the vegetation. Some vegetation is adapted to fire and regenerates very quickly. But if a large fire goes through a boreal forest, it will be destroyed for decades. Fire has the impact of re-starting vegetation or causing the vegetation to shift. It is a very complex chain: climate affects vegetation, vegetation affects fire, fire affects vegetation, and then fire has a direct impact on the climate again. I don't think that we've got a complete picture of these interactions, and there are no models so far that include real fires.

What are the impacts of humans on wildfires? Is it merely the starting of wildfires?
Most people would think of starting fires, whether accidentally or deliberately. That is actually a relatively minor human influence on fires. The suppression of fires by humans is much more important. We fragment the landscape by having crop land and building roads, so fires can't spread as far. If you look at charcoal records of fire over the last 2,000 years, peak fires started going down over the 20th century because people have been fragmenting the landscape more and more. The role of landscape fragmentation also varies with vegetation. If you fragment a savanna by putting in farms, it has no impact on the fire regime because that landscape was already relatively fragmented. If you fragment an area that hosts more continuous vegetation and does not have very many fires, like the boreal forest, you actually increase the number of fires because you're creating pathways where the fires can start and regions where the fuel can dry. If you go to the fire-adapted Mediterranean and do the same fragmentation, you will decrease the fire.

You also studied ecological processes in the ocean with the Dynamic Green Ocean Project. How do plant traits in the ocean differ from terrestrial plant traits?
I think the ocean is much simpler than the terrestrial biosphere, and the traits you need to think about are simpler, the most important one being size. The adaptations of the ocean biosphere are more about migrating over water depth than over distances like terrestrial plants do. But I don't think there is any fundamental difference in the question of adaptation to environmental conditions.

Your Group "SPECIAL" (Sandy's Palaeo Environments and Climate AnaLysis) analyses past climates and creates huge databases. How can those databases help to predict future climates?
The main purpose of those databases is model evaluation and understanding climate processes. One example is the so-called ACER Database, where we put data together looking at rapid climate changes in the past, termed the "Dansgaard-Oeschger events." Those were times of rapid climate warming; in Greenland, we had temperature changes between 5° and 15° in a decade. That is an example of a climate change

as big and rapid as we are expecting over the 21st century. This means that we can look at the processes that generated those climate changes, and we can also look at the impact. One of the things we did was to look at how these warmings in Greenland will be expressed globally. In the northern latitudes, the changes are essentially expressed in terms of temperature changes, whereas in the tropics, they are very much tied in with changes of rainfall in the monsoons. That's the first thing you can do with these big global databases: ask, "What process is going on?"

You could also ask, "Are those processes being captured by the models we're using to predict the future?" That is where the model evaluation comes in. Overall, the models are getting the directions of changes right, but very often they're underestimating the magnitude of changes. If you take those results together, you can say that we're probably underestimating both the variability of climate in the 21st century and the magnitude of change that we're going to get from CO_2 forcing, and that gives you a warning to allow for some safety margin when you're planning how to adapt to climate change.

Do you sometimes feel that by studying proxy and palaeo data, you get lost in the past instead of focusing on the future?

I don't think so. I think a lot of people focus on the future, which is very dangerous because we can't test whether we are right or not. We put a lot of effort into predicting the future, which is really not scientific. What we should be using is our experience of the past and of the present, and we should be cautious about what we need to do to adapt to climate change. I don't believe we are going to stay at 1.5 °C; I don't believe we are going to stay at 2 °C. But we can look at the past and say, "How do we adapt to what must be happening?" We've seen in the past that you can have decade-long droughts, megafires, and so on, and they are going to happen in the future. We don't know when, we don't know exactly how, but they are going to happen. We need to use that experience to shape how we think about the future.

How did vegetation adapt to those Dansgaard-Oeschger events in the past? What can we expect if warming happens?

The vegetation shifted dramatically. In Southern Europe, we shifted from open, very grassy vegetation in the cold periods to tree-covered vegetation in the warm periods. That happened over a period of 20–100 years. The natural vegetation will adapt to those rapid climate changes. When people say, "the velocity of climate change in the 21st century is so fast that the biosphere will not adapt," I think they're fundamentally wrong. The only problem is that people have fragmented the landscape. In the Dansgaard-Oeschger events, there's nothing to stop a tree from migrating. Today, if seeds land in a field, they're going to get ploughed up next year. There is a little bit of limitation to what the natural vegetation can do, but it's not a fundamental limitation. The plants will adapt, that's perfectly clear to me.

Normally, if you think about a tree putting down seeds, it distributes them maybe a kilometre away. If you think about that as being the only mechanism by which they can move, you come up with the calculations that the IPCC did last time round, which is that trees move very, very slowly. But if a bird grabs the seed and flies 20 km before it drops it, or even further, you have these so-called "long distance facilitators of migration." That odd event is still sufficient for trees to move—and we can see that they did. It's those very events which you can't observe where the palaeoclimate provides this information.

How have climate modelling and database construction changed since you began working or researching in this field?
In terms of climate modelling, the models have gotten more and more complex, there are more and more of them, and more and more people are involved in their development. It has become a major industry where every country has its own climate model with a team of 300 or 400 people. When I first started working with climate models, our team was university-based and included maybe half a dozen people.

These big models have become much less transparent and easy to use because they are more complex and more difficult to understand. There has also been a shift in the way we use them. When we first started with climate models, they were very much hypothesis-testing devices. This is what a model is to me. We asked, "Can we explain the expansion of the monsoon through changes in the Earth's orbit—yes or no?" That was the sort of simulation we did. Now the models are basically projection machines. They're expected to tell you the incidence of droughts in 20 years time and so on. The purpose of models has changed fundamentally. It has moved away from being a scientific tool to being a political tool.

In terms of databases, the big thing has been Open Access. When we first started putting data together, there was no demand for Open Access. There was no option to share things except via the post. I remember my first computer, when sending an e-mail message took a long time. You typed the message very slowly, and then a day later it would arrive at the other end. Nowadays, we expect instantaneous responses and the ability to talk to people online. I think that there has been a shift in terms of data accessibility and reproducibility. And that is really good because all too often useful and valuable data was lost because it was in somebody's drawer, and they retired or changed jobs or offices, and the data was lost. The whole process of publishing and doing research has changed. It's not administrative, but there are a lot more steps that people have to do when they're publishing.

The role of the biosphere is often not included in climate models. Do you think that this is changing right now?
Most of the Earth system models we are working with have some representation of the terrestrial biosphere. My complaint about that is that they oversimplify the terrestrial biosphere. They adopted a technique that we introduced when we created the first biogeographic model (BIOME) of dividing the world up into plant functional types. Most of the models have between five to thirteen or so plant functional types, and that is an oversimplification. But at the same time, it's too complex because they require numbers for all of the processes for each of those plant functional types. As we know, there's more variation within a plant's functional type than there is between them. That becomes kind of complicated. What's really missing in the models is wildfire and its interactions with vegetation. There are so-called fire-enabled vegetation models, but they are not used in the Earth system models that predict the climate. Also, plant hydraulics are oversimplified: the way plants take up water and their responses to

moisture changes are not well treated in the models. That is why the LEMONTREE project is exciting, because we are tackling some of these things.

You take part in a lot of collaborations; how important are those for climate research?
I think they're absolutely vital. The days when you could go out as an individual scientist and make a discovery have long passed. Now, we need to have fairly large teams. That is for several reasons. First of all, interdisciplinary research is really crucial for understanding climate change and the Earth system. You can't expect someone to be an expert in plants, animals, climate, and water, as well as in statistics and modelling. Getting teams together that have different expertise and perspectives is really important. If you have big teams, you also stand a better chance of being critical about the premises that you are putting forward. You avoid confirmation bias because there is always somebody who says, "But what if...?" I think that is really important. Being part of large teams is certainly the thing that's made my career in climate change research very interesting but also very productive. I first started off with COHMAP (The Cooperative Holocene Mapping Project), which was about 60 scientists working together. The PMIP (Paleoclimate Modelling Intercomparison Project) is made up of hundreds of scientists who gather to look at issues, and the LEMONTREE (Land Ecosystem Models based On New Theory, obseRvations and ExperimEnts) project, which we've just started, has 50 PIs that are all working very intensely on one topic.

Was there a moment in your career that you found particularly exciting or surprising?
My first encounter with climate models was an eye-opening moment. I was in a position to say that I wanted to be involved in this. That was in the very early days of climate modelling.

How did you overcome obstacles entering the male-dominated area of climate modelling as a female non-physicist?
I was lucky because since my mentor John Kutzbach was about the least sexist man I've ever encountered, that was fine. It's true that, in general, science is still a fairly sexist environment. I think I overcame it by being rude. In this environment, you just have to be pushy.

It's worth remembering that there are other kinds of knowledge besides physics. I think one of the reasons that climate models have generally screwed up in terms of the terrestrial biosphere is that they assume that plants think like physicists—and plants don't.

What are the advantages and disadvantages of working in an interdisciplinary field?
The disadvantage is that some specialists think you don't know anything. I have had situations in the past where people have assumed that I don't know anything about a particular subject, and you have to demonstrate that you do before they let you into the club. But, of course, you do get to see things from different perspectives and

understand more about the way things work. The world itself is interdisciplinary. If you start thinking like that, you get better at understanding how things work.

Which result of your research are you particularly proud of?
That's tough. I'm always in favour of the thing I did last week. If I were going to pick one paper that I am particularly proud of, it would be the summary of our pollen reconstructions for the last glacial maximum and the mid-Holocene. The reason I'm proud of that is because it took several hundred palaeontologists from around the world and probably 15 or 20 meetings to get all that data together. I personally organised, motivated, and shouted at people about handing over data, and I made my postdocs sit up at night typing stuff into a computer from paper sheets. It was a really proud moment when we got that paper together and said, "This is it, it's done."

What development in climate research gives you hope?
I hope we manage to get a better terrestrial biosphere into the climate models, and I hope that we return to using climate models for testing hypotheses and probabilities rather than as a "This is what's going to happen" machine. Those are the things I'd like to see happen in the next three to five years.

How much time did you and your team spend on inventing cool acronyms for your projects? We noticed quite a few nice ones.
I'm not very good at acronyms, so I farm them out to somebody else. SPECIAL was invented by one of my postdocs. LEMONTREE was invented by one of my colleagues in the project; he's very good at coming up with acronyms. So I make use of other people. The logo of LEMONTREE is actually a real lemon tree that is growing in the office of one of the PIs. We invented this acronym for the proposal, she planted a lemon tree, and when we heard back that it was funded, the lemon tree flowered. We are now preserving this lemon tree with the hope that it'll produce fruit within the project's lifetime, and if that happens, there will be a big splash on the web page *(laughs)*.

Would you say planting a tree is a good form of carbon compensation?
(*Laughs*) No, I wouldn't. No. First of all, you can only plant trees where trees will grow. I remember a few years ago, just before the Beijing Olympics, the Chinese had planted trees along hundreds and hundreds of kilometres, all of which were busy dying because it wasn't an environment that would support them. I think that rather than planting trees, people should be thinking about how they use resources. Unless we cut down on our energy use and are more realistic and responsible about how we use resources, anything we try to do is just going to be a minor sticking plaster on what is actually a huge wound. There are other reasons to plant trees—they're nice! Sometimes we use these things as an excuse for not taking action.

Chapter 8
Stephan Weber: "The Urban Body Affects the Local Atmosphere in Many Ways"

Paule Hainz, Lena Hilf, Lukas Schmitt, and Cecilia Lange

©Technische Universität Braunschweig

Stephan Weber (*1974) is Professor of Climatology and Environmental Meteorology at the Technische Universität Braunschweig, Institute of Geoecology. He studied physical geography and climatology at the Ruhr University Bochum. In 2004, he finished his Ph.D. in urban climate research at the University of Duisburg-Essen, where he continued working as a scientific assistant. Today, his research mainly focuses on boundary layer and urban climatology, micrometeorology, and urban air quality.

Date of the Interview: February 3, 2023

G. Lohmann, *Conversations on Climate: The People Behind the Science*, SpringerBriefs in Climate Studies, https://doi.org/10.1007/978-3-032-19147-2_8

To what extent does climate change concern you in your daily life?
Climate change concerns me as a topic, a phenomenon, and also as a problem and challenge. Professionally, it comes up in the context of my teaching. I don't do classical climate change research in the sense of climate system or paleoclimate research, but whenever one deals with questions of change in environmental systems, we also deal with the influence of climate change. I also think about climate change in my private life, asking questions like, "What can I do myself? How can I renovate my apartment to improve its energy efficiency? How can I commute more sustainably?" These are questions that play a role almost every day and, fortunately, have become much more widespread in society in recent years.

Does it help to be able to intellectualize the issue of climate and environmental problems, or does that make it harder sometimes?
It makes it easier in the sense that you know both the processes and the implications. I know that it's not just a topic that currently has media presence and social relevance, but I also know the background. I believe that, in particular, broad social knowledge is still lacking. Even though we've been doing knowledge transfer for many years, I often wonder how well basic difficulties, problems, and processes are understood by the broader population. And I would suspect that it would be much easier to respond to climate change and to take responsibility if one understood the connections better.

What do you think is most lacking in order to make progress in climate action as a society?
I think there is a lack of knowledge transfer and information to be able to act and decide consciously, even if it's just little things. For me, the most obvious aspect is to consider where you can change your own mobility, for example. You should rethink your own attitude: every kilometer that you don't travel in a motor vehicle, after all, helps in the end.

What does your everyday working life look like?
I have the privilege of being able to plan my professional life pretty freely—in theory. Major cornerstones are the courses during the semester, which are firmly scheduled in the timetable. Others are university self-government appointments, such as committee meetings, which take place regularly. The other aspect of my work is research, which is very broad. This includes supervising doctoral students in my research group, theses that are ideally research-oriented, and research as a free space where I myself sit, think about things, and work with data. I think the modern life of a professor has less and less of the latter—really doing things yourself—and more of the former—mentoring researchers. That's not a bad thing; you just have to get used to it. In this respect, regular appointments to discuss interim results, new findings, difficulties, and solutions to problems are part of the weekly program. Although research often takes a long time, sometimes it also happens very quickly: you are inspired within seconds by an article written by other researchers and see how it would advance your own research, for example, by a method presented. Those are always nice moments.

How did you get involved with urban climate?
It started when I was a student. I studied in the Ruhr area, so urban space has always been a topic. During my studies, I had a professor whose research focused on the fields of precipitation climatology and statistical climatology but who also gave a lecture on urban climatology. I found that very interesting at the time, but I also thought that urban climate mainly focused on overheating and the urban heat island, which we already knew quite well, so it didn't seem very appealing to me. Originally, I didn't have a scientific career as a goal; this emerged step by step. After my diploma thesis, I became interested in scientific work and started looking for Ph.D. positions. In the end, I wound up at the University of Essen (now Duisburg-Essen) in a research group for urban climatology. Then I realized that my original impression of the topic of heat islands wasn't wrong at all, but due to urbanization and climate change, this topic has become much more important over the years, also from the perspective of dealing with overheating. This is a much bigger issue in cities than in non-urbanized areas.

What are other research aspects of urban meteorology?
There are many ways in which the urban body affects the local atmosphere as energetic exchange processes between the surface and the atmosphere are modified. The materials used have different thermal properties than natural materials in the surrounding countryside or on natural land surfaces. In addition, there are geometric aspects, such as the three-dimensionality of urban development. These are more indirect effects. Of course, there are also direct effects due to humans, such as heat emissions and emissions of trace substances. In a city of 250,000 residents like Braunschweig, many have a similar daily routine, like commuting to work in the morning. This produces clear patterns and processes, such as peaks in pollutants during the morning rush hour, but we also end up doing many different things, which in turn creates a lot of variability that has to be taken into account. That's what's so exciting about urban climatology.

Can urban climatic effects already be accounted for in large-scale climate models these days?
Yes, that is already being done. It is a big issue at the moment and probably will be in the coming years. For example, you could look at classical weather prediction models that operate at a resolution where individual grid cells span several kilometers. Cities are still mapped in a very rudimentary way, based only on their surface use compared to surrounding areas (sea, land, and so on). However, there are also higher-resolution regional models, which are then driven by the larger models. The higher the spatial resolution, the more urban processes are represented, but still very much through surface use. One must ask whether these typical urban climatic processes are sufficiently resolved in these models. For example, the three-dimensionality of building development still needs to be represented as a subscale process. It's in there somewhere in the grid cell, but we can't resolve it explicitly. The challenge or the exciting thing is—and now I'm leaving my actual expertise—that with increasing computational power for weather forecasting, we are moving towards a kilometer-scale resolution. And at that point, it will be extremely exciting to be able to explicitly

simulate urban climatic or meteorological processes. Hopefully, this will lead to improved weather forecasting in and around cities.

Which effects do cities have on regional climate?
Two important processes are the different heat absorption and release to the atmosphere compared to an undeveloped site. Urban materials absorb more heat, which is then released back into the atmosphere with a delay. This warming and cooling behavior is very different between cities and their surroundings. This is the classic case of the urban heat island. The city remains warmer at night, and in contrast to the surrounding area, it also releases this heat. This is also related to the question of the influence of cities on precipitation. These are processes that can be studied via modeling.

How do cities influence the amount of precipitation?
This has been studied for many years, but there is no single clear answer yet. We know that the amount of precipitation increases in the wind shadow of the city, the so-called leeward. One hypothesis is that convection over the city drives precipitation-triggering processes. There is also evidence that cities can deflect approaching thunderstorms. This is referred to as bifurcation. The effect of cities on the atmosphere continues to be a subject of research.

What is the impact of climate change on cities and on the climate in cities? What is the global influence of cities on climate change?
The majority of people live in cities, and the majority of anthropogenic CO_2 emissions originate from urban areas. That's where we have to start, and there are already many opportunities for more climate-efficient action and climate neutrality.

What kind of adaptation measures can be taken against urban heat waves?
Shading is the absolute key during the day to avoid high radiant temperatures, that is, the energy absorbed by the human body due to solar radiation and radiation from hot surrounding surfaces such as asphalt. A vivid example is when you step out of the blazing sun into the shade. You notice a clear difference. It feels more comfortable, although if you were to hold a thermometer in your hand while doing this, you would hardly measure any difference in air temperature, maybe 0.1–0.2 °C.

What role do plants—especially trees—play in making cities more resilient?
In my opinion, a major one. That's something we've been discussing intensively for a few years now. Trees provide shade while capturing carbon, but with greening, there are always tradeoffs to consider. A tree that provides shade can also obstruct flow and reduce wind speeds. That's why you always have to consider where you want to plant. On a busy road, this would not necessarily be beneficial, as the trees provide a reduction in ventilation. Attention should also be paid to the spacing at which trees are planted and the canopy shapes that allow better air exchange. Overall, in addition to providing shade, vegetation brings the advantage of transpiration. Accordingly, less energy is used to directly heat the air and more is used for evapotranspiration, which contributes to cooling. Of course, as drought conditions increase, the limited water supply becomes a problem. It is also unclear how the measures will be accepted

by the population. At a science night once, there were many people in favor of more trees, as long as they didn't drop sticky sap on their cars. I was very surprised at how often we received this feedback. It reminded me that you have to take public opinion into account and be aware of conflicting interests. I think it's important that positive developments in the urban climate don't get a negative connotation.

How much attention does urban climate research get in local politics?
As far as I can tell, there is definitely much more attention than 15 or 20 years ago. In my early days, there was often criticism that the research results were seen far too little. I would assess that a bit differently nowadays. But it depends very much on the local situation and whether there are people in city planning offices and environmental offices who have the topic on their radar. There are some cities that are very advanced and innovative in this respect and are happy to pick up on these ideas. For example, Braunschweig is on the right track.

We've already talked about greening. What influence do green roofs have on the urban climate?
Green roofs can evaporate water, contribute to local cooling, provide habitat for insects, and serve as insulation against high temperatures. Thus, they also benefit the climate inside the building. These effects are well known. In our research, we are particularly interested in the extent to which the exchange of heat, water vapor, and CO_2 can be quantified using modern measurement techniques. My impression is that the research community is partly operating with rudimentarily measured values, which were obtained in overly brief measurement campaigns with very simple approaches. Therefore, we have been working on a methodical approach to quantifying these exchange processes for several years. For this purpose, we consider longer time series and focus strongly on CO_2 exchange. In particular, we are investigating whether a green roof can be a sink for CO_2 by taking up more CO_2 through photosynthesis than it releases through respiration. We have been able to obtain very positive and robust results: green roofs absorb CO_2 over many years. This also happens during dry periods, although the uptake is then reduced by almost half. A comparison has shown that green roofs are able to absorb about half the amount of CO_2 that lawn sites can absorb. There are quite a few unused roof areas that, in the worst case, absorb a lot of energy due to their dark color and contribute to the overheating of cities. There is still a lot of unused potential. Extensive green roofs are relatively barren, thin substrates planted with drought-resistant species. These substrates have little organic material and therefore exhale relatively little CO_2, yet these drought-tolerant species are able to take up CO_2, and that's a pretty good compromise.

What makes your measurement approach for CO_2 exchange different from previous ones, except that the time series are longer?
In the air layer near the ground, the exchange processes between the surface and the atmosphere are turbulent. The method we use has been around for 20–25 years. It measures mixing in the atmosphere and the transport of heat and CO_2 with high time resolution, which means that this turbulent flux can be determined directly and, unlike other approaches, no assumptions are made about the turbulent state of the

atmosphere. However, this requires a large area. We are currently measuring at two sites; one is about 9,000 m^2 and the other is 70,000 m^2. A roof area of 200 m^2 is not sufficient for turbulent flux measurements because the space next to the roof would disturb the measurement.

There is a lot of talk about the necessity of using roofs for solar panels in cities, but if green roofs also play a role in mitigation and transformation, what should be prioritized?

In most cases, a square meter of green roof can absorb less CO_2 than the same area of photovoltaics can by replacing other types of energy generation. However, there may be roof areas that are suitable for greening but less so for photovoltaics. So I wouldn't play them against each other. It's currently a big topic to combine photovoltaics and agricultural land. Especially for the combination of photovoltaics with green roofs, there are still too few studies. However, rather positive synergy effects are to be expected since green roofs lead to lower surface and near-surface temperatures. This would improve the efficiency of photovoltaic cells. That's where a study looking at exchange processes, heat fluxes, and CO_2 fluxes would be useful and is also on our agenda. The difficulty with this is the large roof area needed for the method we are using. At the moment, we have no knowledge of a roof area with such a combination.

How well can green roofs be applied to existing structures?

To the best of my knowledge, retrofitting a roof with extensive greenery is usually not a problem from a structural point of view and is not too expensive. In Germany, the majority of green roofs are extensive. Intensive roof greening, on the other hand, means creating entire roof gardens. However, these have more demands on the substrate, which must be 30–50 cm thick instead of 10 cm. The structural requirements are correspondingly different and should be planned from the outset.

Can you give a rough estimate of the percentage of houses where retrofitting would be possible?

In recent years, I have repeatedly tried to obtain reliable figures on how many roofs in Germany or in individual cities are greened. Little is known about this other than the annual increase. This is calculated indirectly via the amount of substrate that the green roof manufacturers apply. Attempts have been made to determine the existing green roof area via remote sensing. Green roof registers exist for some German cities. From this, it can be deduced that there are cities with 1–2 m^2 of green roof per inhabitant. However, this is still a small proportion and thus offers great potential.

Would green roofs also have a positive influence on the urban climate in drier regions than Germany?

Yes, because they are optimized—in terms of the type of greening—exactly for this climatic region, insofar as they are capable of surviving and growing. Plants on roofs are often not sown over the whole area, so the amount of vegetation still multiplies over time. One problem is a sufficient water supply so that evaporation can take place. Measurements at the Berlin site have shown that a dried-out green roof does not behave much differently thermally than an unvegetated roof. With thin substrates, this can happen quickly. On drier sites, you would have to think about irrigation, such as collecting precipitation and feeding it to the green roof.

You're also doing research on cold air. Why is this particularly important in cities?
Cold air refers to an air mass that is cooler relative to the urban atmosphere. Over certain surfaces, air can cool much faster in the evening or at night, when solar radiation is no longer present. Earlier, we had the example that urban building materials such as asphalt and concrete are slow to release heat back into the atmosphere and thus remain warmer than, say, lawns. Thus, colder air masses accumulate over unsealed surfaces. We are interested in the situations in which these air masses start to move. This depends, for example, on differences in air pressure, which then initiate compensating flows. Colder air masses, typically from the outskirts of the city, start moving towards the city and can contribute to local cooling. This works even better when this cold air movement is driven by terrain gradients; this is because cold air, which can be thought of as a tough, pulpy mass that flows almost like honey, is denser than warm air and starts moving by gravity as soon as it is formed on sloping terrain. There are, for example, the so-called Höllentäler valleys near Freiburg, in the Black Forest. There, large volumes of cold air can form and move towards the center of Freiburg. In cities like Braunschweig, which do not have large topographical differences, the flow is due to temperature differences. This cold air movement can penetrate to the edges of the built-up area but is severely impeded by vertical obstacles such as buildings. If one wants to enhance the cooling of a city, one needs a certain openness in the form of aisles through which the cold air can penetrate.

Another problem in cities is fine particles. How should one imagine the path of particles through city air?
Particulate matter is defined as particles smaller than 10 μm (PM_{10}) and is usually expressed as a mass concentration. We are more concerned with ultrafine particles, which are the smallest kind of particulate matter. These are particles smaller than 100 nm. In the measured PM_{10}, they represent only a very small part of the mass but are present in very high numbers. Environmental medicine says that precisely these abundant, small particles have significant effects on human health because we absorb them very deeply when breathing. Thus far, there is too little data available to be able to make clear proposals regarding limit values. Combustion processes, which include industrial processes and motor vehicle traffic in particular, are the dominant sources of these particles in cities. Diesel particulate filters have brought some positive changes, but there is still a relatively high level of exposure to ultrafine particles. We are interested in what happens to them upon their emission. In the past years, our research has focused on the question of where we find significantly enhanced concentrations and how these particles are transformed during atmospheric transport.

Where do these ultrafine particles accumulate in cities?
We see that traffic is still a dominant source. The closer we get to a major road, the higher the concentration becomes. In the so-called urban background—for example,

within residential areas with low-traffic roads, in green areas, or in parks—the concentrations drop significantly.

Are there more unusual sources of ultrafine particles?
Yes, for example, inland navigation on the Rhine in Cologne and Düsseldorf. There is a lot of tourist boat traffic there—a source that has not been studied thoroughly. Ship emissions have been studied intensively in recent years, but mainly in the nearshore area. Mobile sources such as construction machinery and construction work are generally dominant but difficult to monitor. Either you have a monitoring station that continuously collects data and temporarily shows the influence of mobile machinery, or you would have to directly monitor the machinery in some way, but in the latter case, there is no long-term view to put this in context.

At the beginning of the COVID-19 pandemic, we saw that traffic decreased rapidly. Could this be seen in the measurements of particulate matter?
Yes. There are many publications on air pollutants as well as greenhouse gas concentrations. The difficulty is that concentrations are dependent on weather conditions. You have to put a lot of effort into figuring out how the period of the first major lockdown differs from the periods before and after. Here in Germany, weather conditions changed significantly when the lockdown began. When compared to other times, these weather changes need to be taken into account. There have been many studies using various methods to filter out this weather effect. One could see a reduction due to the mobility change ranging from a few percentage points to as much as 40%. Another neat way to do this is to look at data from exchange measurements. With these turbulent flux measurements, you look at a particular part of the Earth's surface to quantify the direct exchange. Specifically, it determines how much airborne matter—be it pollutant or greenhouse gas—is transported per area and time. Since we measure the transport rate and not the concentration, the background from weather change is not critical. There is a study that performed such measurements for CO_2 in various European cities and found reductions of up to 80%. We studied this for ultrafine particles in Berlin and also showed a significant reduction of about 38%. In general, the COVID-19 pandemic was very interesting for air quality studies because it was a big, unforeseeable experiment.

Are there special challenges when taking measurements in cities?
There are very practical, pragmatic challenges, such as vandalism. I myself have had few difficulties so far. Once a measuring station was broken into, but the thief apparently realized that the content was of no use to them. Nothing further happened. Another challenge is to isolate the effects of certain events from measurements. There is always a lot of heterogeneity in a city; it is alive and pulsating. For example, if you plant trees or a hedge along a street and want to know whether it helps filter air pollutants, you have to think very carefully about how cleanly you can demonstrate this effect. For example, changing wind directions have to be taken into account. Of course, it's also an exciting challenge to see if new ideas or methods can be used to investigate new, previously unresolved effects.

Is there a research result that you are particularly proud of?
There are two things from the recent past, although that doesn't mean that other results aren't important, too. All members of my group are enthusiastic researchers with lots of passion for their data and their different topics. One important topic is green roof research, especially because it grew out of a fairly small idea. What was originally a thesis idea has grown into a nine-year data series. Since dry, hot years were included in this period, we can now answer further questions. We are now much more interested in the green roof as an ecosystem than we were originally. In general, there are institutions that focus their research on measuring long-time series, but there are also many groups that work on a more project-based basis and only take measurements over weeks or months. Experience has shown how important long-term measurements are for our understanding of the system. I am now much more aware of the potential of long-term measurements in new projects. This also applies to the second field of particles and particle exchange. Through long-term measurements at a site in Berlin, we have gained insights that would not have been possible over a short period of time.

What developments in climate research give you hope?
I am very hopeful that many researchers are becoming more aware of the value of high-quality fundamental research. There is also greater consideration of how this can help generate socially relevant knowledge and products. A current example is the measurement of CO_2 exchange: "What possibilities are there to simplify and expand this measurement so that it can be carried out at more sites? Can we standardize this process and thus make a statement about whether a site is a CO_2 source or sink?" From this, we could derive climate protection issues, such as how to deal with ecosystems in order to maintain or optimize this sink function. We now have large amounts of data that can help us achieve this goal. There is a trend toward making such data available for broader use instead of remaining on university servers. I think this is an encouraging development.

Chapter 9
Friederike Otto: "The Paris Agreement Is a Human Rights Treaty"

Karolin Stiller, Marius Schulz, Ulrike Richter, and Julius Mex

©David Fisher

Friederike Otto (*1982) is a climatologist, physicist, philosopher, and leading researcher in the field of attribution science. She is primarily concerned with the influence of human-made climate change on extreme weather events and their impact on society. Otto studied physics at Free University of Berlin and received her Ph.D. in the epistemology of climate models. Following her Ph.D., she worked at the Environmental Change Institute at the University of Oxford for ten years. Since 2021 she has been a Senior Lecturer at the Grantham Institute for Climate Change and the Environment at Imperial College London. She is co-founder of the international

---

Date of the Interview: November 28, 2022

G. Lohmann, *Conversations on Climate: The People Behind the Science*, SpringerBriefs in Climate Studies, https://doi.org/10.1007/978-3-032-19147-2_9

World Weather Attribution project and one of the lead authors of the IPCC Sixth Assessment Report.

What do you particularly enjoy about your work?
I think it's harder to find something that I don't enjoy. My work is exciting, intellectually challenging, and relevant to our society. We are looking at extreme weather events not because it's easy to look at them in a climate model but because something has happened in the real world and people have questions about it. That may not be roller-coaster fun, but it's the kind of fun you want to have at work. A large part of this is that I've made the decision to only work with people who are fun to work with.

When did you make that decision?
It's a very conscious decision of mine that I don't work with assholes anymore. Before I came here to Imperial College, I was at Oxford for ten years. When I started there as a postdoc, people would often say to me, "Don't you want to write a proposal together with him or with her? That would be good, wouldn't it?" And today I would say "No!" This doesn't always work out 100%, but I think otherwise it would be hard to endure the academic world.

When you think back to your student days, were you hoping to end up where you are now?
No, I never did any career planning in that sense. Actually, I didn't really want to go into research or science.

Did you have a vision at that time?
Honestly, not really. I knew that I definitely wanted to study and really get into a subject, but I never really had a vision of what I wanted to do with my degree. I started studying physics and then began a second degree in philosophy, in which I also wrote my doctoral thesis. In terms of career planning, that was about the stupidest thing I could have done because it meant that I was not at all employable in the academic field in Germany. Actually, I wanted to do science communication afterwards, but I didn't have the experience for that. So, climate research was not at all what I always wanted to do. I couldn't think of anything better, and then I happened to see this postdoctoral position at Oxford.

Did you have role models or people who supported you along the way?
In Germany, definitely not. Going to Oxford was my idea. I applied for a job there, which I didn't get at first, but my interviewer referred me to his colleague, Prof. Myles Allen, who had just had a postdoc position become available, and basically told me to do whatever I wanted to. So, I started to look around to see what I could do and ended up deciding to work on a project where a climate model is simulated not on a supercomputer but on the computers of regular people who provide their computing power. This way, extremely large ensembles can be calculated—for instance, thousands of realisations of possible weather scenarios instead of only ten, or nowadays maybe hundred. Therefore, this model is predestined to be applied to extreme weather.

Did you have moments of doubt along the way about how to proceed?
I had a pretty bad Abitur (high school diploma), which is why I couldn't study any subjects with entrance requirements. So physics was the lesser of the evils. My undergraduate degree, in particular, was not a lot of fun. Above all, it was difficult. It wasn't until my graduate studies that I discovered philosophy as a minor subject, which I did enjoy. My doctoral thesis was also very frustrating. It was really only my dog that saved me from going insane.

In your doctoral thesis, you worked on how climate models can be used to gain insights into the climate system.
Exactly, whereas I originally wanted to pursue the question of what can be learned in principle from models from an epistemological point of view and what this means for the communication of scientific results. The fact that my thesis was then mainly about climate models was due to the fact that I had worked with a climate model at the Potsdam Institute for Climate Impact Research (PIK). A doctoral thesis never turns out the way you imagine it at the beginning. That's the most important thing you have to accept as a doctoral student.

Are your philosophy studies still relevant to your current work?
Yes, definitely! Philosophy is about questioning the assumptions behind every step. I think that's what actually enabled me to become a good scientist. It gets you out of the nitty-gritty of methods and questions and makes you question what you're doing yourself, what other people are doing, and what you could do differently. I didn't learn scientific work or how to question things at all in my physics studies.

With your philosophical background, how would you describe the relationship between climate models and the real climate system?
That depends entirely on which climate models you look at and which questions you want to answer. There's this nice saying by John Box, "All models are wrong, but some are useful." It's a question of whether the available models can answer the specific question you have. There is no such thing as the perfect model. For example, if you want to look at global precipitation changes, there are models that are good at representing certain aspects of the circulation over Europe and others that are good at doing so over the monsoon regions. An important basic rule is to never only use one model.

What are the models that you work with?
Two very different types of models. Most of the climate models that I work with are general circulation models (GCMs). These are models that have a complex representation of the atmosphere, ocean, etc. These models were developed by the major climate computing centres and are also used by the IPCC. In addition, I still work with statistical models. Basically, this is extreme value statistics, which are based on observational and modeled data.

How would you describe your relationship to climate models?
Climate models are insanely useful. They are based on the same principles as weather prediction models. The weather is not given only by energy and mass conservation;

these two just provide a rough framework. Additionally, conservation of momentum and the chaotic nature of weather are important. The statistics of these dynamics over long periods of time then yield the climate data. With the climate models, you can determine causalities that you couldn't get at all from purely observational data.

When you work with a model, do you have a concrete idea of the model, or do you just work with the output?
I don't know. I know the equations that the models are based on, and I know what they mean. In addition, of course, I know what the input of the models is. In this regard, it's not just the output, no, but thank God I don't write the code for the models myself.

Could you briefly outline how attribution research works methodologically?
The idea behind it is actually not very difficult. You compare the weather in the world we live in today—that is, a world that has warmed by 1.2°—with the weather of a world that is just like our world today but without human-made climate change. In doing so, one finds differences in how often certain weather events occur. This can be attributed to human-made climate change. This is attribution research.

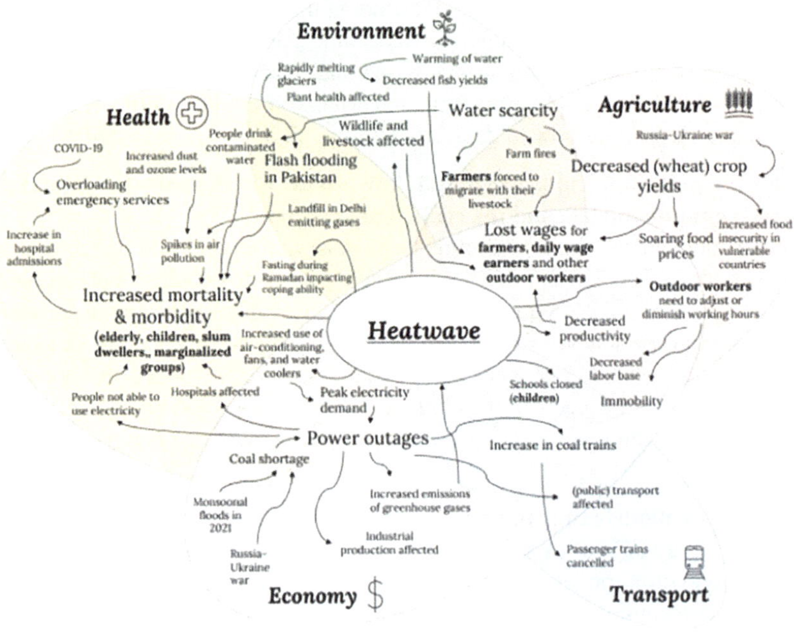

Conceptual map of impact pathways during a heatwave. Figure from Zachariah et al. (2022).

We asked you to bring your favourite graphic from your research. Could you briefly describe what it's about?
The figure addresses the fact that the chance of extreme weather events becoming humanitarian disasters depends only to a very small degree on climate change and the weather. In fact, it mainly depends on vulnerability and exposure, people and ecosystems that are specifically exposed to the extreme weather event. The conceptual map represents how heat waves affect social systems globally and locally. That's what's almost always ignored in the climate change debate. When people talk about climate change, it's usually about physical things like global temperatures, potential tipping points, or other disaster scenarios. The Paris Agreement, however, is a human rights treaty, not a treaty to save polar bears or humanity from extinction. The point is that changes in weather and climate are increasing inequality around the world, curtailing basic human rights such as the right to live. This is not talked about enough, but that's exactly why we care about climate change in the first place, exactly why it's a problem. At the same time, it also shows all our possibilities. It's not an asteroid falling on our heads, to which we are completely defenseless. Instead, it's something we can do quite a lot about on many different levels—especially when it comes to extreme weather events. I mean, mitigation and ending the burning of fossil fuels are important aspects, but vulnerability and inequality also need to be reduced. That's why I think this graphic is so important.

Where does the general public overestimate or underestimate the explanatory power of climate models or attribution research?
They are actually not overestimated. After all, we've spent the last 50–60 years listening to the fossil fuel industry say that the uncertainties in climate science are way too large. The only thing where they might be overestimated are local heat waves. We always think heat waves are the easy thing—the thing that is really well understood. In fact, the changes in the observational data for extreme heat at the local scale are very different from the data that the models give. However, because the general perception is that heat waves are well understood, few people actually do research on them.

In what areas are models often underestimated, then?
It's annoying to hear, "Oh, that's loaded with uncertainties," as if that's something special or as if that in some way discredits what's being said. All science is subject to uncertainty. Another problem with existing climate models is that they have all been developed in the Global North, not one in Africa. They are always developed for a specific region of interest. The UK Met Office climate model is really good for the UK climate, but that's just a pretty small island in a pretty big world. Political decisions and the North-South discrepancy in science contribute to climate models being much worse at modelling tropical climate. What's more, this has something to do with the climate itself. Unlike the mid-latitudes, where the climate is based on statistics, the tropical climate is much more influenced by large-scale phenomena like El Niño. But if you don't develop a climate model yourself, then you have no choice but to use the models that are there, even if they contain major deficiencies with respect to tropical climates.

You co-founded the World Weather Attribution. What's the scientific mission or motivation behind this?

When extreme weather events take place, the media and policymakers always immediately ask, "Is this climate change?" But that in itself is a poorly posed question that cannot be answered with yes or no. It's more about what role climate change plays in this event. For a long time, all sorts of people had opinions and answers to that question, except scientists. They always said that nothing could be said about individual weather events. We've changed that with attribution research. However, the time scales on which scientific publications take place are usually such that a study is not published until two years after the event took place, and by then the public discussion has moved on. We founded World Weather Attribution to do it faster, on time scales where the media or decision-makers are still interested in the results. So, attribution was born out of frustration that scientists never said anything publicly about the role of climate change in weather events.

Do you also think that there is a scientific mandate to deliver these findings?

There is a total justification for blue-sky thinking in science for things that have no immediate relevance, but there is also a mandate for science to deal with issues that are relevant right now. I do think that science should deal with the real world!

What were some particular challenges you faced both when founding World Weather Attribution and advancing it?

It's just a lot of work to do it so quickly. It's about bringing high-quality scientific evidence for the role of climate change within two weeks with the expectation that the results will stand up in a peer review process. Normally, that would take six months. That only works if you have a team with enough people who know what they're doing. Afterwards, almost all of our studies have gone through the peer review process to ensure that our methods are robust and that our numbers stand up to scrutiny. In the beginning, we didn't publish much because we were mainly developing methods. When we finally sent our first rapid attribution study for peer review, it was rejected—not because the scientists actually had issues with the method but based on the reasoning of "That's too fast, science doesn't work that fast. *Reject.*" Later, with new data and different models, we did the exact same study again and got exactly the same result. Then it made it through the peer review process without any problems.

The peer review process is the Holy Grail of science, and for good reason. Therefore, the biggest challenge initially was the criticism from our own scientific community. After all, it wasn't just about doing these studies but also about talking about them and reaching journalists and the media. If you do science and nobody knows about it, you can't influence the debate, but that has changed by now. Of course, there are still some scientists who say, "Ohhh, this would never go through peer review." Then you wait six months and you can show that yes, it does.

What are the limits or obstacles that you face in your current work?

For years, we had no funding at all. The World Weather Attribution was pure goodwill from colleagues around the world who thought the project was important and collaborated with us. That has the side effect of my network being quite large, and I work with colleagues from many different countries around the world. Nowadays, there is some funding, but by far the biggest challenge remains people power, because the thing that takes time is the scientific input at the beginning of a study when it comes to finding out what happened in the first place. For example, we might get a notification about food insecurity in West Africa from the Red Cross. Then you have to figure out which aspect of the weather actually led to these impacts. Was it the lack of or delay in the onset of the rainy season? Did the cause even take place in the area where the impacts are now? Or is it actually because of other mediating factors? You can't automate that. Once you make those definitions, the actual attribution with weather and model data is relatively straightforward. But when it comes back to interpreting the results—which models to trust—that's another step that requires intellectual input. That's the bottleneck.

How do you envision the field of attribution research in the future?
I would very much like the national weather services to become more active themselves. They always refer to our studies and are enthusiastic when we ask them to collaborate. However, except for the UK Met Office, the German Weather Service, and maybe Météo France, the national weather services don't have the know-how and capacity to do so. If they take over the attribution studies for frequent events like heat waves or heavy precipitation, we could focus on complicated things, such as the interplay of weather, climate, vulnerability, and exposure. That would be my ideal, but that's still quite far away.

Do you have a result from your research that you're particularly proud of?
I am very proud of what World Weather Attribution is now and what we have achieved. I mean, half of Chapter 11 of the last IPCC report is based on studies that we've done over the last eight years. Of course, there are studies that I think are better and studies that I think are worse. Mostly, it's the current studies that I find good and the older ones less so, which is because you're always learning and improving.

You have already touched on the connection between extreme weather events and society. What potential does attribution research have there?
I think it's incredibly important that we understand what climate change means in concrete terms. It's not about abstract predictions from climate models, it's about how climate change manifests itself and what impacts it has on our society. In large part, these impacts consist of changes in the frequency of extreme weather events and how these in turn interact with our social systems. I think it's important to connect local, social, and socio-economic experiences with intellectual knowledge about climate change. Simply to understand that it is not about polar bears or the end of the world but about our social structure. On the other hand, attribution research methods work not only retrospectively to say, "This is how much the weather has already changed," but also to analyse how it will change in the future. That is crucial information to base policy adaptation measures on.

In addition, attribution research closes a gap in the causal chain. We know exactly who emitted how much greenhouse gas and how much a given amount of it changes the global mean temperature. Additionally, you have the damage caused by an altered occurrence of extreme events. Through attribution research, one can now trace the direct causal chain from the countries' and companies' emissions to the damage, and the responsibilities become clear. This can become crucial, for example, in court.

You just mentioned the issue of climate justice. Would you like other researchers in your field to be more publicly political?
Well, I wish that we would no longer have a debate about whether science is political or not. Of course science is political. There is this idea that both the media and natural scientists still adhere to that science is somehow neutral, value-free, or objective. It is not. Every decision in science—what questions you research or who gets how much funding—is always political. Still, that doesn't mean that political attitudes influence the results.

To illustrate this with an example, in 2021, the World Food Program called the famine in Madagascar the first famine in the world attributable to climate change. We did an attribution study on that and found that climate change didn't actually play a role in the specific drought that led to this famine. The region just has very high climate variability over long periods of time. There are always years with lots of or little rain. At the same time, there are many social reasons why a crisis occurs as soon as there is less rain. It would have been nice if we had found out, "Yes, climate change plays a big role." That could have drawn a lot of attention to what climate change specifically means. But our results did not confirm it in this case. This does not mean that climate change is irrelevant in Madagascar, but that for this one famine, climate change was not critical. Of course, we also studied this drought because we want to provide scientific results that support campaigns such as those of the World Food Program. Unfortunately, that doesn't always work out, but that is exactly the crucial point. We studied it for political motives, but that does not make the results biased.

You've expressed thoughts in public posts about how we can stop reproducing post-colonial power structures. How would you like to see this dealt with by the scientific community of the Global North to avoid this reproduction?
An important step is indeed the realisation that today's science engages in exactly this type of reproduction. Only what White men in the Global North consider to be science is considered scientific. Science is based on peer review, and peer review is done by those who do science now. As a result, other methods and approaches are often simply rejected—not because they are wrong, but because they do not conform to what the European-American scientific tradition has established as science. Many scientists—more male than female scientists, but not only—are simply not yet aware that this is a problem. It would be important that people who study science also learn what the biases are. That is, how and in which structures science is done. That doesn't fix the problem immediately, but it would at least make it smaller, especially for future generations.

And once you have that awareness—
Yes, awareness alone is not enough, although it is an important prerequisite. There are also very concrete things. For example, it costs money to publish scientific studies. Most scientific journals are paid by scientists to publish. This means that organisations with little money and with scientists who are not linked to universities and therefore do not have the money to do so cannot publish in top-tier journals. Either you would have to abolish publication fees altogether or subsidise them, for example, by a country that wants to provide development aid.

Another thing would be something that is always discussed under the buzzword "capacity building." Most women scientists in the Global South don't have a mandate to do science but are hired to provide a service. This can include weather, seasonal forecasting or teaching, but as a full-time job. In Germany, there are all kinds of institutions like Helmholtz and Fraunhofer societies that allow research as a profession. That doesn't really exist at all in the countries of the Global South. Building and supporting such capacities would be a very important, concrete step that could start today.

Who is responsible for initiating these changes?
Different levels. On the one hand, countries like Germany or the UK provide development aid, but this is usually not designed for the long-term social or societal change that would actually be necessary to really fight inequality. Instead, this money usually flows into easy-to-plan, short-term aid. There needs to be a rethink in the ministries of the Global North. This also applies to institutions like the IPCC. For example, when selecting scientists, the IPCC is now placing more emphasis on having more scientists from the Global South, but that's where it ends so far. This does not change the fact that they actually do not have time to contribute to the IPCC because of their already mentioned full-time employment. It would help if the IPCC would finance their work, allowing them to actually participate in writing the global climate report. Another concrete thing would be for the IPCC to invest in decent Internet in the universities where they work so that the WiFi is stable enough that they can participate in online meetings without interruptions. In that sense, there are scientists in the Global South who are co-authoring IPCCs but are left on their own. You can be as brilliant, capable, and knowledgeable as you want and still fail because of these very practical barriers that are actually easy to overcome.

You've already touched on the fact that what is practised as science depends on what is accepted as science. If we imagine a world without these patriarchal structures, what would feminist climate science look like? What value could such a world add to science?
I don't know what that looks like in concrete terms because I studied physics in Germany, where I was mostly the only woman. That has very much shaped my image of science. But I think one problem is that quite legitimate questions at the interface between social and physical problems are hardly ever dealt with. Even in relatively interdisciplinary climate research, you still have the natural scientists on one side and the social scientists on the other. There is still too little communication and collaboration between them. Additionally, the social sciences are still seen as

less relevant than anything that can somehow be expressed in numbers. In the future, that must no longer be the case.

What would have to change in order to perceive these kinds of questions and research projects as legitimate?
I don't know how to really do it, except to just try. In my research, I have never gotten research money from government institutions for what I wanted to do. Most people who write reviews for research projects are older scientists close to retirement. They have time to do this because they no longer have to worry about their own careers. But that means that whether funding takes place or not is still heavily influenced by people's disciplinary stereotypes. That really needs to change.

In your chapter *Unlearn Science* in the book *Unlearn Patriarchy*, you introduced the term anti-patriarchy buddies. Can you explain that briefly and tell us who that is for you?
There are a lot of things you can't do alone. If you want to restructure a department or make science more participatory, you need fellow women. You also often don't see your own biases and need someone to point them out. The idea of these anti-patriarchy buddies is to lead to more voices being heard and thus science actually reflecting the real world. It is important that one not be in a dependent relationship with these buddies so that they can criticise you. Besides, the task of these anti-patriarchy buddies is simply to do science and to do it better. The two most important buddies for me are Paola Arias, a professor at the University of Antioquia in Colombia, and Joyce Kimutai, Chief Scientist at the Kenya Meteorological Service. Then there are others from the IPCC or who I work with from time to time.

How did you meet the two women you mentioned?
I met Paola at the IPCC writing the Sixth Assessment Report. Through that experience, we learned to fight together. In the Summary for Policymakers, there is an illustration that consists of hexagons. Paola and I designed this figure together and fought hard to get it into the report. We actually had to get our colleagues to agree to this, and, unusually, the governments agreed immediately. That really bonded us a lot. After that, we both got the same hexagon-shaped earrings—I almost put them on today.

Joyce and I met through a project with the Red Cross in Kenya in 2015 or 2016. She was involved as a representative of the Kenyan weather service. When we met, she found attribution and our project so exciting that she joined some World Weather Attribution studies. Right now, we are doing a project on heat waves in Kenya. In the meantime, she is the Focal Point for Kenya in the IPCC, which means that she is also part of the government delegation that is present at the Conference of the Parties (COP). Besides our exchange, we also talk a lot about issues like loss and damage.

In the same book (*Unlearn Patriarchy*), you wrote, "Only if you know the unwritten, patriarchal rules can you examine, unlearn, and actively change them." What rules do you find most difficult to fight and unlearn?
Well, what I still find very difficult is that men always seem more intelligent to me, for example, when I supervise Ph.D. students. Men have to do less than women to

give me the impression that they are smart. That's a bias I still have very much inside me, even in job interviews. It's sad, but yeah...

We've already talked a lot about desirable future changes. What are your dreams for the next ten years? What are some exciting questions that you want to deal with?
Actually, I would like to retire in ten years and only deal with exciting things, but that's not happening. I think that just in the next few years, a lot will happen in climate litigation and lawsuits. The idea is to sue for climate justice not only on a political level but with the help of the courts. That's where I started working a lot with lawyers.

What developments in climate research give you particular hope?
I think that, at least in part, the boundaries between disciplines are dissolving. The natural, social, and human sciences are working more and more together on adaptation research.

What advice would you like to give to people who are just getting into science or thinking about it?
Don't work with assholes! I didn't dare to say until quite late, "Nah, I only work with the people I want to work with and who are fun to work with." I think that not only increases your own quality of life but also lets you achieve a lot more scientifically. For example, when it comes to what Ph.D. you want to do, I don't think it's so important whether that's at a prestigious university or with prestigious people, but what the work-life balance is like. There are so many academic institutions and academics for whom it is completely normal to work on weekends or constantly late into the night. They expect the same from you. If you're not in the office at least until 6 PM or even later, you are given the side-eye. There, people are proud if they have worked the whole weekend. Not going along with that and learning to say "no" is the most important thing. It's not easy, and it's a long way to achieve that.

Reference

Zachariah M et al (2023) Attribution of 2022 early-spring heatwave in India and Pakistan to climate change: lessons in assessing vulnerability and preparedness in reducing impacts. Environ. Res.: Climate 2 045005. https://doi.org/10.1088/2752-5295/acf4b6

Glossary

Atlantic meridional overturning circulation (AMOC) The Atlantic Meridional Overturning Circulation (AMOC) is a large system of ocean currents that carry warm water northward into the North Atlantic Ocean. This results in a milder climate in northwest Europe and the North Atlantic, relative to the Pacific Ocean. The AMOC is maintained by dense water sinking in the North Atlantic, from where it flows back south along the North American east coast. Increasing ocean temperatures and a greater influx of freshwater from melting glaciers may thus lead to a weakening of the AMOC. The near-surface flow of warm water from the Gulf of Mexico, which is part of the AMOC, is also called the Gulf Stream (Good et al. 2018; Fofonoff 1981; Ackermann et al. 2020).

Differential equations Differential equations are mathematical equations whose solutions are not numbers but mathematical functions. They are the way most physical laws are expressed since they can be used to model how a given state of a system changes with time or in space. Oftentimes, a given physical quantity also depends on other quantities or their evolution. For example, pressure, density, and flow velocity in a fluid depend on each other and cannot be determined independently. In such a case, multiple differential equations are said to be coupled. Due to the complexity of the equations required to study, say, the evolution of the climate, computers are heavily relied on to solve them, which is referred to as numerical treatment of differential equations (Dahmen and Reusken 2008; Landau and Lifshitz 1987).

El Niño El Niño is a climate pattern that recurs every 2–7 years. During El Niño years, the ocean surface temperature in the eastern tropical Pacific warms significantly. This induces the formation of heavy rainfalls in South and Central America and influences climate patterns all over the globe (McPhaden et al. 2020).

Entropy and the second law of thermodynamics Entropy is one of the most important quantities in physics, especially in thermodynamics. From a thermodynamics perspective, changes in a system's entropy are closely related to the transfer of heat energy into that system. Entropy is essential in understanding how much of a system's internal energy can be used to perform work, for example, in heat engines. The higher the entropy of a system, the less free energy is available to

G. Lohmann, *Conversations on Climate: The People Behind the Science*, SpringerBriefs in Climate Studies, https://doi.org/10.1007/978-3-032-19147-2

extract work. Sunlight has low entropy and ultimately enables many processes on the Earth, such as ocean circulation, which are vital to the evolution of life. Microscopically, entropy is closely connected to disorder. For example, consider a box of gas particles that is divided into two halves, A and B. Assume all particles start out in A. Left to themselves, the particles tend to distribute equally between both halves, thereby increasing the disorder in the system and, hence, its entropy. This observation is very general. The second law of thermodynamics states that the entropy of an isolated system cannot decrease over time. Fittingly, the arrow of time (from past to future) can be physically defined through entropy (Nolting 2017).

Grids A subdivision of the Earth's surface into smaller cells to solve climate models. Numerical climate models operate on increasingly finer grid sizes as the performance of parallelized super computers increases. Whether a model can represent a geophysical process depends on the model formulation and discretization. Since the spatial scale of oceanic eddies is 1–100 km, the ocean model grid resolution needs to be on the same scale to represent these ubiquitous small-scale features. Alternatively, their effects must be parameterized as in current state-of-the-art general circulation models, where the typical horizontal resolutions of the ocean component is about 1°. Practically, given the present high-performance computer capacities, efficient and parallelized model codes, it is now possible to conduct simulations for 50–100 model years per day (Climate Models 2023; Lohmann et al. 2020).

Heat engine A heat engine transforms heat energy into mechanical work. For example, a car combusts fuel and uses the resulting heat to ultimately move forward. The second law of thermodynamics restricts the efficiency of this process: by converting heat energy to work, the automobile's entropy is lowered. At the same time, the entropy of the surrounding environment needs to increase. Otherwise, the second law is violated. This is why not all heat energy can be converted into work. In the case of a car, excess heat with high entropy is blown out of its exhaust pipe. The concept of a heat engine can also be applied to the Earth system. Sunlight heats up the atmosphere, which leads to atmospheric motion, or wind. The maximum possible efficiency a heat engine can achieve is known as Carnot efficiency (Nolting 2017).

IPCC and Summary for Policymakers (SPM) The Intergovernmental Panel on Climate Change (IPCC) is a multilateral body established by the World Meteorological Organization (WMO) and the United Nations Environment Program (UNEP) in 1988 to share the state of knowledge about climate change. It consists of 195 member states, which elect a bureau of scientists to lead the preparation of the IPCC's reports for a cycle of 6–7 years. Three working groups and a taskforce are responsible for examining all relevant scientific literature on climate change, including natural, economic, and social aspects. Hence, the panel does not conduct scientific research itself but rather produces an objective and comprehensive overview of the current state of knowledge compiled into publicly available assessment reports. The findings of these reports are reviewed by thousands of leading scientists and endorsed by all member governments. The Summary for

Policymakers (SPM) aggregates the findings of the IPCC's assessment reports mainly for policymakers but also for the general public. Other than in the main report, every line in the SPM is negotiated and needs to be approved by the governments of the member states to ensure a balance between accuracy, clarity of message, and policy relevance (Solomon et al. 2007).

Jet stream Jet streams are fast-moving, narrow and meandering air currents located near the top of the troposphere at an altitude of about 7–10 km. They are susceptible to small perturbations flowing from west to east, leading to huge eddies with storms and fronts. Jet streams are generated by the temperature gradient between the equator and the poles. The most important jet streams are the polar jet stream and the subtropical jet stream (at 30° and 60° N/S latitude). Nowadays, commercial jets fly at an altitude close to the jet stream. This is why flights from west to east generally take less time than flights from east to west (Woollings 2020).

MM5 Starting in the 1960s as a model for monsoons, MM5 became, in the 1970s, a broadly used model capable of simulating many atmospheric phenomena, real-time forecasting, and climate studies on the mesoscale. MM5 is short for Fifth-Generation Penn State/NCAR Mesoscale Model (Anthes 2011).

Monte Carlo simulations Monte Carlo methods were first developed to solve certain mathematical problems (integrals) in many dimensions, where techniques such as evaluation on discrete grids become infeasible due to the exponential growth of the number of grid points needed with the number of dimensions. They approximate the true solution by statistically sampling the parameter space according to a distribution that generates large probability mass in areas of importance while vanishing in areas of low importance, i.e. where the integrand is vanishingly small. This technique is referred to as importance sampling. The distribution with which the parameter space is sampled must be chosen to fit the problem and does not necessarily generalise from one problem to another. This statistical approach is where this class of methods derives its name from, a tongue-in-cheek reference to the location of a famous casino by the same name. (Kalos and Whitlock 2008).

Navier-Stokes equations The Navier-Stokes equations are a set of partial differential equations that describe the motion of fluids by considering factors like pressure, viscosity, and velocity. They consist of the continuity equation, the momentum equation, and the energy equation and, in most cases, require numerical methods to be solved. Climate models use the Navier-Stokes equations and other physical laws to simulate the fluid motion of the atmosphere and ocean. However, these equations are highly non-linear and computationally demanding (Thorne and Blandford 2017).

Overland flow Overland flow is water that, instead of, for example, infiltrating into the ground, flows on the surface of the terrain (Ward and Robinson 2000).

Paleoclimatology Paleoclimatology is concerned with reconstructing the states of the Earth's climate in periods in which direct measurements did not yet exist. Paleoclimatologists obtain data from a variety of archives, including rocks, sediments, boreholes, ice sheets, tree rings, corals, shells, and microfossils, and apply a variety of different techniques to date the information. This allows a deeper understanding of the evolution of the past climate, such as glaciations, rapid cooling, and warming events. The knowledge gathered by paleoclimatologists also provides an important basis for the study of past changes in the environment and biodiversity, specifically the relation between climate and mass extinctions as well as biotic recovery (Bradley 2015; Lohmann 2009).

Proxy Data In paleoclimatology, or the study of past climates, scientists use indirect information to reconstruct past climate conditions. This data documents characteristics of the environment that can stand in for direct measurements. Paleoclimatologists gather proxy data from natural recorders of climate variability such as tree rings, ice cores, fossil pollen, ocean sediments, corals, and historical data. By analyzing records, scientists can extend our understanding of climate beyond the instrumental record (What are proxy data? 2018).

Stomata Stomata are gap openings in the epidermis (protective outer layer) of plants. They are used for gas exchange, mainly transpiration and CO_2 uptake, and can be opened or closed by the plant via guard cells (Bresinsky et al. 2013).

Uncertainties in scientific results The word "uncertainty" has slightly different meanings when used in everyday speech versus a scientific context. In scientific discourse, it conveys the degree to which something is known. In the vernacular, the word conveys a sense of not knowing. The difference is subtle but important. Uncertainties in scientific experiments or models make probabilistic statements about the interval in which the "true" value of a variable is expected to be found. These uncertainties can arise from various factors, such as the variability of experimental measurements, numerical errors caused by approximations made in computer models, and the inadequacy of the underlying physical model of a problem. One should therefore not confuse the careful scientific practice of error estimation with a lack of conclusive results, as the former does not in general imply the latter (ISO 1995; Kennedy and O'Hagan 2001; Scientific uncertainty 2019).

References

Ackermann L et al (2020) AMOC recovery in a multicentennial scenario using a coupled atmosphere-ocean-ice sheet model. Geophys Res Lett 47(16). https://doi.org/10.1029/2019GL086810

Anthes R (2011) History of the mesomonster. In: Tom Warner symposium, 2 Dec 2011. Accessed 02 June 2024

Bradley RS (2015) Paleoclimatology: reconstructing climates of the quaternary. Elsevier, Amsterdam. https://doi.org/10.1029/EO081i050p00613-01

Bresinsky A et al (2013) Strasburger's plant sciences. Springer, Berlin, Heidelberg. https://doi.org/10.1007/978-3-642-15518-5

Climate Models (2023). https://www.climate.gov/maps-data/climate-data-primer/predicting-climate/climate-models. Accessed 07 Sept 2023

Dahmen W, Reusken A (2008) Numerik fur Ingenieure und Naturwissenschaftler. Springer, Berlin, Heidelberg. https://doi.org/10.1007/978-3-540-76493-9

Fofonoff NP (1981) The Gulf stream. In: Evolution of physical oceanography: scientific surveys in honor of Henry Stommel, pp 112–139. https://doi.org/10.1029/EO063i019p00499-02

Good P et al (2018) Recent progress in understanding climate thresholds: ice sheets, the Atlantic meridional overturning circulation, tropical forests and responses to ocean acidification. Prog Phys Geogr Earth Environ 42(1):24–60. https://doi.org/10.1177/0309133317751843

ISO (1995) Guide to the expression of uncertainty in measurement. Technical report, Geneva, pp 16–17. https://doi.org/10.1017/CBO9781139135085.018

Kalos MH, Whitlock PA (2008) Monte Carlo methods. Wiley-VCH, Weinheim. https://doi.org/10.1002/9783527626212

Kennedy MC, O'Hagan A (2001) Bayesian calibration of computer models. J R Stat Soc Ser B: Stat Methodol 63(3):425–464. https://doi.org/10.1111/1467-9868.00294

Landau LD, Lifshitz EM (1987) Fluid mechanics. Course of theoretical physics, vol. 6. Pergamon Press. https://doi.org/10.1016/C2013-0-03799-1

Lohmann G (2009) Abrupt climate change modeling. Springer, New York, pp 1–30. https://doi.org/10.1007/978-1-4419-7695-6_1

Lohmann G et al (2020) Abrupt climate and weather changes across time scales. Paleoceanogr Paleoclimatol 35(9).https://doi.org/10.1029/2019PA003782

McPhaden MJ, Santoso A, Cai W (eds) (2020) El Nino southern oscillation in a changing climate. Wiley, Geophysical monograph series. https://doi.org/10.1002/9781119548164

Nolting W (2017) Theoretical physics 5: Thermodynamics. Springer, Cham. https://doi.org/10.1007/978-3-319-47910-1

Scientific uncertainty (2019). Nat Clim Change 9(11):797. https://doi.org/10.1038/s41558-019-0627-1

G. Lohmann, *Conversations on Climate: The People Behind the Science*,
SpringerBriefs in Climate Studies, https://doi.org/10.1007/978-3-032-19147-2

Solomon S et al (2007) IPCC, 2007: summary for policymakers. In: Climate change 2007: the physical science basis. Contribution of Working Group I to the fourth assessment report of the Intergovernmental Panel on Climate Change. Cambridge University Press, Cambridge and New York, NY

Thorne KS, Blandford RD (2017) Modern classical physics: optics, fluids, plasmas, elasticity, relativity, and statistical physics. Princeton University Press, Princeton, 1511 pp. https://doi.org/10.1080/00107514.2018.1515249

Ward RC, Robinson M (2000) Principles of hydrology. McGrawHill, London. https://doi.org/10.1002/esp.3290160308

What are proxy data? (2018). https://www.ncei.noaa.gov/news/what-are-proxy-data. Accessed 07 Nov 2023

Woollings T (2020) Jet stream: a journey through our changing climate. Oxford University Press. https://doi.org/10.1093/oso/9780198828518.001.0001

GPSR Compliance

The European Union's (EU) General Product Safety Regulation (GPSR) is a set of rules that requires consumer products to be safe and our obligations to ensure this.

If you have any concerns about our products, you can contact us on ProductSafety@springernature.com

In case Publisher is established outside the EU, the EU authorized representative is:

Springer Nature Customer Service Center GmbH
Europaplatz 3
69115 Heidelberg, Germany

Batch number: 10492512

Printed by Printforce, the Netherlands